T0340792

Total Volunteer Force

The Hoover Institution gratefully acknowledges
THE SMITH RICHARDSON FOUNDATION
for their significant support of this publication.

Total Volunteer Force

LESSONS FROM THE US MILITARY ON LEADERSHIP CULTURE & TALENT MANAGEMENT

Tim Kane

HOOVER INSTITUTION PRESS
STANFORD UNIVERSITY | STANFORD, CALIFORNIA

With its eminent scholars and world-renowned library and archives, the Hoover Institution seeks to improve the human condition by advancing ideas that promote economic opportunity and prosperity, while securing and safeguarding peace for America and all mankind. The views expressed in its publications are entirely those of the authors and do not necessarily reflect the views of the staff, officers, or Board of Overseers of the Hoover Institution.

www.hoover.org

Hoover Institution Press Publication No. 681
Hoover Institution at Leland Stanford Junior University,
Stanford, California 94305-6003

Copyright © 2017 by the Board of Trustees of the
Leland Stanford Junior University

All rights reserved. No part of this publication may be reproduced, stored in a retrieval system, or transmitted in any form or by any means, electronic, mechanical, photocopying, recording, or otherwise, without written permission of the publisher and copyright holders.

For permission to reuse material from *Total Volunteer Force: Lessons from the US Military on Leadership Culture and Talent Management*, by Tim Kane, ISBN 978-0-8179-2075-3, please access www.copyright.com or contact the Copyright Clearance Center, Inc. (CCC), 222 Rosewood Drive, Danvers, MA 01923, 978-750-8400.CCC is a not-for-profit organization that provides licenses and registration for a variety of uses.

First printing 2017

23   22   21   20   19   18   17   9   8   7   6   5   4   3   2   1

Manufactured in the United States of America

The paper used in this publication meets the minimum requirements of the American National Standard for Information Sciences—Permanence of Paper for Printed Library Materials, ANSI/NISO Z39.48-1992.♾

Cataloging-in-Publication Data is available from the Library of Congress.
ISBN-13: 978-0-8179-2075-3 (pbk : alk. paper)
ISBN-13: 978-0-8179-2076-0 (EPUB)
ISBN-13: 978-0-8179-2077-7 (mobipocket)
ISBN-13: 978-0-8179-2078-4 (PDF)

Too often, our military is losing and misusing talent because of an archaic military personnel system. Promotions are handed out according to predictable schedules with only secondary consideration of merit. That's why even after more than a decade of service, there is essentially no difference in rank among officers of the same age. . . . We should ask whether we should give commanders greater discretion to build a staff with the specialists and experts they need in the right positions. Commanders are likely better able to assess their needs than bureaucrats in the personnel system.

—SENATOR JOHN McCAIN, CHAIRMAN OF THE SENATE ARMED SERVICES COMMITTEE, DECEMBER 2, 2015

It is time to reevaluate whether the Defense Officer Personnel Management System, commonly referred to as DOPMA, continues to meet the needs of our military services.

—SENATOR JACK REED, RANKING MEMBER OF THE SENATE ARMED SERVICES COMMITTEE, DECEMBER 2, 2015

We stand in line. We wait for our number to come up at a board, and you're either in or you're out. And so it's an up-or-out system, which, again, has worked very well for us, but I see that changing over time.
. . .
The accelerant in this is Congress, because they write the law. DOPMA and the law and personnel policy are authorities granted to us from Congress, so they have to take this on as much as we want it.

—VICE ADMIRAL WILLIAM MORAN, CHIEF OF NAVY PERSONNEL, DECEMBER 9, 2014

Dedicated to the memory of my friend Mike Tomai.

Thank you for volunteering.

Contents

Foreword

Over forty years ago, Milton Friedman and his friend Martin Anderson put forward the idea of ending the draft and recruiting volunteers for the armed forces. Since the draft ended in 1973, the concept of the volunteer armed forces can now be said, unequivocally, to have succeeded. The nation has been privileged to have talented young men and women volunteering to serve in our armed forces and maintaining our national security.

The ironic thing about the All-Volunteer Force is that those who enter the armed services volunteer only once—when they join. After joining the armed forces, their careers become subject to a variety of regulations, regardless of their own preferences. Commanders and service chiefs have little control either, because the law now requires that assignments are made by centralized bureaucracies and that promotion timelines are identical for each service. Isn't it about time that someone took a look at this situation and made some suggestions?

Well, someone has. Tim Kane has thought about this issue and has provided us with a blueprint of how our volunteer forces can be improved. This is a worthwhile endeavor because it will undoubtedly enhance the quality, readiness, and efficiency of the work done by our armed forces, whether in combat or in manning the posts that help secure our borders and our interests around the world.

Tim Kane has done us a service in writing this book, and his work leads to other issues that must be addressed. One of the problems of the Pentagon's current personnel system is that pensions and health care costs for retirees are growing at a rate that, if nothing is done, will require a huge portion of the defense budget that is needed to support our national security. It is clear that the health care system is in need of reform, and the growing use of health savings accounts is one step in the right direction. In this volume, Kane effectively takes on important aspects of the pension problem. Replacement of the current pension system of defined benefits with one of defined contributions, in which individuals match government contributions, would remove the pension liability overhang.

This change is also notable because it would help preserve the concept of volunteerism. The current system almost ensures that an individual in the armed forces will stay for twenty years and, often, not a day longer, because that is when the pension system kicks in. Under the Kane system (increasingly used by private US employers), there is no such turning-point date, so a spirit of volunteerism and its benefits can continue indefinitely. He has other suggestions for compensation reform that are worthwhile as well, rooted in ideas that many other economists have written about for quite some time, notably our fellow Hoover Institution colleagues Eric Hanushek and the late Martin Anderson, who was himself the godfather of the 1973 reforms. Incentives matter. And the military has for too long incentivized seniority at the expense of merit, talent, and skills for the twenty-first-century security environment.

GEORGE SHULTZ
Thomas W. and Susan B. Ford Distinguished Fellow,
  Hoover Institution, Stanford University
US Secretary of State (1982–89)
Captain, US Marine Corps (1942–45)

Acknowledgments

First and foremost, I would like to thank the men and women who serve in the US military, including individuals on active duty, those in the reserves and National Guard, and veterans. Hundreds have given their time and wisdom to this project, and I hope the result is worthy of their respect.

Literally hundreds of senior officers and government officials have advised me and provided feedback on drafts, and I am especially grateful for informative visits and opportunities to discuss the research during sessions at the Pentagon, Naval Personnel Command, US Air Force Academy, University of Kansas, the Philanthropy Roundtable, and the Naval Postgraduate School.

This book was made possible by a generous grant from the Smith Richardson Foundation, and I am grateful for its support.

Thanks to the Hoover Institution, which has given me a home for the past three years to conduct this research. I am especially grateful to my Hoover colleagues, including George Shultz, Condoleezza Rice, Gary Roughead, Jim Mattis, David Slayton, Joe Felter, John Taylor, Eric Hanushek, Eddie Lazear, John Cochrane, David Henderson, Josh Rauh, Amy Zegart, Richard Epstein, Kori Schake, Chris Dauer, Stephen Langlois, Mike Franc, John Raisian, and Tom Gilligan. In addition, I received unexpected and generous input from Charles O'Reilly, a professor at Stanford's Graduate School of Business.

His impact on my thinking and on much of the text was tremendous. A great deal of thanks goes to four classes of the Hoover Institution's National Security Affairs Fellows, a program generously supported by Bob Oster and other Hoover Overseers who have also encouraged this work. Finally, I must acknowledge the inspiring work done by the late Hoover scholars who paved the way for all of us in promoting an All-Volunteer Force: Milton Friedman, Gary Becker, and Martin Anderson above all.

Research support from Evelyn Heil was invaluable. Thanks also to Chris Dellaporta, Shane Culver, Alexis Gurganius, Alexa Liautaud, and Laura Olson for their dedicated assistance. Many other colleagues at Hoover generously assisted the project, and Claudia Hubbard was especially helpful.

Walt Ulmer, Troy Thomas, John Gallagher, Philip Carter, Todd Harrison, Mackenzie Eaglen, Jim Carafano, Kate Kidder, David Barno, Seth Cropsey, Chris DeMuth, Don Vandergriff, Nora Bensahel, and Steve Leonard each had a hand in shaping my thinking. Thanks as well to the many individuals from private-sector firms who provided valuable thoughts, including William Treseder and his colleagues at BMNT partners, Karen Courington, Chris Hsu, John Maguire, Beau Laskey, Joe Wolf, Don Faul, Nathan Estruth, Pat Johnson, Bill Casebeer, Jeff Philippart, Joe Deane, and Steve Kiser. Special thanks to Kevin McConnell and his colleagues at Palantir.

Editors at the Hoover Institution Press have been an invaluable asset in making this book come to life. Thank you for your patience and understanding.

I cannot express enough thanks to Tom Ricks for his encouragement and support in publishing various essays along the way at *Foreign Policy*, and to an equally supportive hand from John Podhoretz at *Commentary*, Ryan Evans at WarOnTheRocks.com, and the editors at *Military Times*.

Although I need to express thanks to so many people, the reader must know that the final analysis and recommendations are mine alone. Indeed, many individuals listed here will and have argued to the contrary on parts of my thinking. Any errors, especially, belong to me.

TIM KANE
Research Fellow, Hoover Institution, Stanford University
Captain, US Air Force (1990–95)

Introduction

The life or death of any organization hinges on the quality of its people, a truth undiminished and perhaps amplified in our era of accelerating technological change. Some organizations such as the armed forces seem immune from catastrophic failure because they are shielded from profits and the risk of bankruptcy, but the stakes are even greater there. The need for human excellence, for trust in teammates, for optimum leadership as well as talent, is universal. And yet how to thrive in the human dimension remains shrouded in mystery.

Two of the most recent defense secretaries have identified the Pentagon's underperforming personnel bureaucracy as the top challenge hindering military commanders and the men and women in uniform. In his farewell speeches and memoir, Robert M. Gates, who was appointed secretary of defense by President George W. Bush in 2006 and reappointed by President Barack Obama, asked how the Army "can break up the institutional concrete, its bureaucratic rigidity in its assignments and promotion processes, in order to retain, challenge, and inspire its best, brightest, and most battle-tested young officers to lead the service in the future?"[1] Ash Carter highlighted recruiting and retaining talent as his top priority during his first speech as secretary of defense, even naming the effort the "force of the future."[2]

The military's dysfunctional personnel system is well known to service members, but it became a national flashpoint during the wars in Iraq and Afghanistan.[3] A number of recent studies have revealed the depth of frustration among active-duty Army officers at all levels. Fewer than one in five officers think the Army does a good job "matching talents with jobs" and "weeding out the weakest leaders," whereas just one in twenty think it does a good job "retaining the best leaders." As alarming as such studies have been, they reveal symptoms of the problem—a point of widespread agreement—but not underlying causes and solutions.

This book endeavors to, first, identify the underlying causes of personnel dysfunction in the US armed forces and, second, propose a set of reforms.

The proposed blueprint aims to move Pentagon personnel policies further along the spectrum of volunteerism, away from the coercive structures that have outlived their purpose after the All-Volunteer Force (AVF) was implemented in 1973. The Total Volunteer Force (TVF) envisioned here would emphasize greater individual agency during all stages of a US military career, not just the first day of enlistment. More importantly, by decentralizing personnel processes, the TVF will restore command authority to colonels and captains that has been missing since the misguided centralization of the 1960s and that has neutered operational flexibility for the past half century.

TVF reforms are fundamentally aimed at getting the right people in the right jobs, which is a job-matching optimization process that decentralized markets do exceptionally well. Service chiefs would have new authority to change promotion timetables (or not), allow greater specialization, allow lateral reentry of veterans to active duty, and increase the hiring authority granted to unit commanders. Improving performance evaluations is another TVF recommendation, which can enhance individual morale and development, but is essential information for improved job-matching. Thirdly, TVF recommendations aim to improve military compensation, which is rife with costly disincentives. These are not recommendations rooted in for-profit best practices; rather, they stem from first principles and an empirical assessment (in chapter 1).

When the AVF was proposed in the 1960s, it was called mercenary by almost the entire defense establishment. President Nixon established a fifteen-member commission to study the issue after the 1968 election, chaired by former secretary of defense Thomas Gates Jr. (no relation to Robert Gates). What became known as the Gates Commission was intentionally balanced at Nixon's insistence, with five members opposed to conscription (including economists Alan Greenspan and Milton Friedman), five neutral, and five in favor. Gates even told Nixon he was "opposed to the whole idea of a volunteer force,"[4] which is why the president wanted him. Many months later, the group formally reported that it "unanimously believes that the nation's interest will be better served by an all-volunteer force. . . ."[5] Now, forty years later, almost no one in uniform wants to serve with draftees. Experience has shown the men and women in uniform to be more professional than ever.

The AVF was a radical transformation that has been fully vindicated by the test of time. Despite deep suspicion among many military leaders and despite a rough first decade of implementation, nobody seriously believes today that a return to conscription would enhance the quality of military personnel. Nor does anyone argue that a draft—*even with lower base pay*—would save money because it's understood that retention would collapse, forcing higher recruiting and training expenses.

The AVF enhanced the quality of the force and its morale and efficiency. Once base pay was raised in the early 1980s to provide competitive salaries, the volume of young citizens willing to volunteer exceeded the number of positions. This led to a steady quality revolution in the ranks to such an extent that now literacy and educational levels of enlistees far surpass average civilian levels, and experienced troops are routinely headhunted by private-sector firms. Despite the dramatic change in how personnel were managed, the AVF transformation should not be considered an end state.

The TVF will extend the core values of volunteerism and professionalism and, like the AVF, will save money. Gains in efficiency should not increase costs, not even in the short term. That is why this book focuses its recommendations on reformed (and restored) processes and authorities rather than on expensive new programs. Reducing coercive personnel processes will reduce compensation needed to push people around.

This is possible because the AVF did not end the use of coercion in the ranks—it only ended it at the accessions gate. Since 1973, coercion has remained the dominant management technique for military HR. Personnel are given orders and their careers are managed centrally, rather than personally. Compensation has been shaped to reinforce coercive control. Most plainly, the twenty-year cliff pension enhances retention with the crudest financial tool. To give a twenty-year veteran a retirement package while a nineteen-year veteran gets nothing would be illegal if the employer were in the private sector, a disparity that strikes most everyone as deeply unfair.

For all the critiques of the personnel system, there has not been much in the way of an alternative. To be sure, reams of Government Accountability Office reports, Rand studies, and think tank papers have highlighted flaws and proposed modifications. In many of the internal government reports, however, the solution to an overwhelmed bureaucracy is bluntly to add more funding to the bureaucracy. As an example, the Army recently moved its Washington, DC–based Personnel Command (PERSCOM) to a sparkling new headquarters at Fort Knox, Kentucky: the newly christened 4,000-strong Human Resources Command (HRC).

The Total Volunteer Force aims to restore command authority over personnel to the US armed forces, much of which was surrendered to centralized bureaucracies during Secretary of Defense Robert McNamara's tenure and codified in subsequent legislation and regulation. Unit commanders today have very little authority to select, dismiss, replace, promote, or rearrange the personnel under their command. Likewise, individuals in uniform have very little authority over their own career assignments. In short, they are not volunteers after day one.

Skeptics of decentralization will rightly warn of nepotism and bias. What should be understood up front is that the theme of these recommendations is flexibility. The armed services operate under extreme mandates, notably the strict promotion timetables that force each service to manage year-group cohorts in lockstep while forcing officers and enlistees into extremely narrow career paths dotted with vital checkpoints and constant geographic and job rotations. The armed forces deserve flexibility to move away from extreme centralization, but the goal is to offer them options and balance. Rather than deal with a binary choice between anarchy and control, an optimal personnel structure should avoid extremes of central or local authority. What the TVF represents is a rebalancing of command authority with central guidance.

The reformer's central dilemma is how to improve the military without sacrificing invaluable traditions. That dilemma can be easily resolved if the reforms are not mandates at all, but simply structured as the removal of stifling mandates in current law and a restoration of service chief and commander authority over personnel policy.

CHAPTER ONE

Analyzing the Problem

How can the Army break up the institutional concrete, its bureaucratic rigidity in its assignments and promotion processes, in order to retain, challenge, and inspire its best, brightest, and most battle-tested young officers to lead the service in the future?

—SECRETARY OF DEFENSE ROBERT M. GATES'S SPEECH AT
THE UNITED STATES MILITARY ACADEMY, WEST POINT,
NEW YORK, FEBRUARY 25, 2011

What is wrong with the Pentagon's personnel system? Perverse incentives in compensation and retirement have distorted the shape of the force—matching highly talented people with the wrong jobs, incessantly rotating employees up and sideways, and fostering a culture where employees feel obligated to express insincere preferences to stay on the career track to "get to twenty." Neutered command authority over personnel decisions makes it difficult to match the right people with the right jobs, hurts readiness, and prevents toxic and predatory individuals from being weeded out of the ranks. Inflated performance evaluations are corrosive to fairness and integrity in the Army, Navy, and, especially, the Air Force.

During his farewell address to the cadets at West Point, Secretary of Defense Robert M. Gates identified the personnel system as his main worry for the future of the Army. Likewise, Secretary of Defense Ash Carter emphasized fixing personnel policies as vital to building "the force of the future" in his first speech in March 2015. Carter said, "We also have to look at ways to promote people, but not on just when they joined, but even more based on their performance and their talent. And we need to be on the cutting edge of evaluating performance. . . . We also need to use twenty-first-century technologies—like LinkedIn kinds of thing—to help develop twenty-first-century leaders and give our people even more flexibility and choice in deciding their next job—in the military."[6]

Red Alert

One statistic above all else serves as a red alert that the military personnel system is dysfunctional: the unemployment rate of young veterans. It averaged nearly a third higher than nonveterans (10.7 percent compared to 8.0 percent) before the 2009 recession. After 2009, more than one in five veterans age eighteen to twenty-four could not find a job between 2009 and 2012, twice the jobless rate of nonveterans. The persistence of this employment gap has reinforced some misperceptions about the quality of troops, even among top policymakers.

The unemployment rate of veterans may seem irrelevant to the readiness of the active-duty force, but it is a profoundly relevant *symptom* of the real problem: the institutional inefficiency of central planning. In blunt terms, some of the nation's most talented young men and women are on active duty but never empowered to take—and in fact are discouraged from taking—an *active* role in applying their unique skills to the military's needs. Job-matching is centrally planned in all of the armed forces. Young veterans enter the private sector almost totally unprepared to search for a job because that activity—the engine that drives America's free market economy—is anathema to the modern Pentagon bureaucracy. It wasn't always this way.

Ben Bernanke, the former chairman of the Federal Reserve, remarked at a public event in August 2015, "If you go into the military at age eighteen—versus an identical person who stays in the private sector and takes a private sector job—ten years later, if you leave the military, your skills and wages are probably not going to be quite as high on average as the private sector person." Bernanke chided the Pentagon for advertising that service in uniform adds beneficial skills: "The evidence appears to be that there really is not an advantage," and further that the military "is really not adding much to the private sector through training or experience." He mentioned academic research by MIT economists Joshua Angrist and Stacey Chen, but unfortunately, Bernanke misinterpreted their work. Angrist and Chen compare veterans to nonveterans from the late 1960s when soldiers were drafted. In fact, the authors conclude that "lifetime earnings consequences of conscription . . . have almost surely been negative." The authors were expressly not analyzing the impact of military service, let alone service in the modern era, but focused exclusively on the negative impact of conscription nearly six decades ago.

So, Bernanke was wrong, but his belief was rooted in the "civ-mil" employment gap that still persists today.

In August 2015, the US Bureau of Labor Statistics (BLS) issued a major report titled *Employment Situation of Veterans*.[7] That month, the national civilian unemployment rate was 5.1 percent. For all 21.2 million veterans, the average rate of unemployment was lower, but for the 3.2 million recent veterans who served during the post-9/11 era, unemployment was higher than the civilian norm. Individuals in this cohort are described as "Gulf War II era" veterans (including all who served after 9/11 as their most recent period of service).

Those who served in Iraq or Afghanistan had lower unemployment rates than other veteran peers, 4.1 and 4.0 percent, respectively. Surprisingly, veterans who had served tours of duty in *both* Iraq and Afghanistan had the lowest unemployment rate of all, 2.9 percent.[8] This suggests that combat experience involves skills that do transition well to civilian jobs. It seems that discipline, courage, teamwork, and other soft skills are highly valued and valuable in civilian occupations.

Another sign of the positive impact of military service comes from a recent study by the Department of Veterans Affairs, which found

TABLE 1.1.
Unemployment Rates of Male US Citizens, by Age and Service (2015 annual averages)

	Nonveterans	All Veterans	Gulf War II–Era Veterans
Total, 18 years and over	5.3	4.5	5.7
18 to 24 years	12.0	13.6	13.6
25 to 34 years	5.4	6.9	6.8
35 to 44 years	4.0	3.8	3.8
45 to 54 years	3.7	3.4	2.6
55 to 64 years	3.8	4.7	4.3
65 years and over	3.6	4.0	-

Source: US Bureau of Labor Statistics

that "post-9/11 veterans attain 11 percent higher median earnings than non-veterans with similar demographic characteristics," an advantage that was even higher for female veterans.[9]

General comparisons of veterans to civilians can easily be skewed by the heavy gender and age differences among those two populations. A closer look at demographically similar cohorts offers an insight into where the problem lies. Note in table 1.1 that younger male veterans have unemployment rates one and a half points higher than male nonveterans, but older veterans have unemployment rates equal or significantly lower.

Why does the employment gap persist for young veterans? The best answer came from a 2014 Rand study by David S. Loughran titled, simply, "Why Is Veteran Unemployment So High?" He considered five different explanations. After extensive demographic analysis, Loughran discovered that the skills mismatch hypothesis—the one Bernanke echoed—has "little support in the available data." Another theory blames service-related injuries, but it is also not supported in the data. Service-related disabilities do keep one in seven veterans out of the labor force, but labor force participation rates are routinely higher, not lower, than for civilians. Nor is employer discrimination to blame. The fourth unsubstantiated hypothesis is that ex-soldiers are less innately competent than other Americans. The opposite is true. Soldiers have much higher literacy and IQ scores than civilians, on average.

Loughran concludes the culprit is weak job search capabilities. Young veterans, by definition, leave a stable job and enter what is to them a strange new world. The key evidence is that the unemployment difference between veterans and nonveterans evaporates over time. It decreases by almost half a percentage point each month after an individual leaves active duty.

In short, US military veterans have superior job skills but no job *search* skills.

Exacerbating the problem are legislative remedies made with the best of intentions. During out-processing, soldiers are strongly encouraged to sign up for unemployment compensation during their first day as a civilian. Yet academic studies show that unemployment insurance (paying people half their recent salary for many weeks if they remain unemployed) raises the national unemployment rate, while also causing skills to atrophy. A smarter program would make jobless benefits more generous, but not allow people to access those benefits until out of work for a month or two. During the 2009 recession, Congress extended the normal twenty-four weeks of jobless benefits to an unprecedented ninety-nine weeks. Young veteran unemployment skyrocketed.

None of this lets the Pentagon off the hook. It is unrealistic to teach a ten-year enlistee how civilian labor markets work with three days of transition classwork. Ultimately, the Pentagon's centralized control over personnel assignments bears the most responsibility. A system of central job-matching leaves soldiers and sailors unprepared for a labor market that requires self-motivation, initiative, and personal responsibility.

Identifying the Real Problem(s)

Symptoms of problems are informative, but should not be mistaken for problems themselves, which is why conversations about retention problems or surveys of morale are of limited practical use. Even broad agreement that personnel policies are dysfunctional is followed by disagreement about what, exactly, is broken and even more disagreement about how to fix things. Respected scholars at the Army War College, West Point, and elsewhere describe the Pentagon system—a one-size-fits-all inflexible set of regulations that binds each branch— as industrial-era or feudal in nature.

The fact that two of the most recent secretaries of defense prioritized personnel issues is a reflection of long-time frustrations in the lower ranks. "The management system created in 1947 to serve a draft military is falling behind the demands of the 21st century all-volunteer force," wrote Marine veteran Jesse Sloman in the *National Interest.* "Critics cite problems throughout the services, including: lockstep promotions based almost entirely on a person's time in service; an outdated method of matching personnel with assignments that does not sufficiently take into account individual preferences, special skills, or unique experiences; and narrowly defined career trajectories."[10]

Citing reformers from high to low, however, could be done at nearly any time going back to the 1950s, when President Eisenhower was calling for personnel system reform. It is an evergreen challenge to modernize the human dimension, just as the military is incessantly challenging itself to remain dominant in weapons technology, tactics, and logistics. What this chapter presents is a more comprehensive assessment in order to move beyond symptomatic and anecdotal descriptions.

This peculiar management dilemma is poorly served by traditional management scholarship. Unlike the typical business school case study, top leaders in uniform do not lack the qualities that are championed by management experts,[11] such as vision, passion, and values. Based on recent studies of the unique paradoxes of military talent management[12] and statements by high-ranking military leaders,[13] the dominant narrative is that the armed forces are headed by superior leaders who feel unable to manage talent effectively because of bureaucratic and legal constraints, despite the military's strong and ancient leadership culture.

Because the military's leadership culture is intertwined with personnel rules, both of which are based on obedience to authority unlike anything in the private sector, there is a real concern that fixing one might irreparably ruin the other. How can personnel (a.k.a. talent management or human resource) policies be distinguished from leadership culture? Which pieces are most important for retention, productivity, and morale? The management literature offers no simple answers for the armed forces, but utilizing firm-level survey data in the manner of Nicholas Bloom offers a way to think about the problem.[14]

The name of the instrument developed for this broad-spectrum analysis is the Leader/Talent matrix, which I designed to evaluate two dimensions of an organization: leadership culture and talent management. The matrix includes forty elements spread across five leadership categories and five management categories. Categories in the cultural dimension are independence, development, purpose, values, and adaptability, which contrast with talent management categories such as training, job-matching, promotions, compensation, and evaluations. A survey of 360 current and veteran service members was conducted in early 2015 using the Leader/Talent matrix. Figure 1.1 shows an overview of survey results by category.

The 360 active and veteran respondents who evaluated the US armed forces gave high marks to leadership culture and low marks to

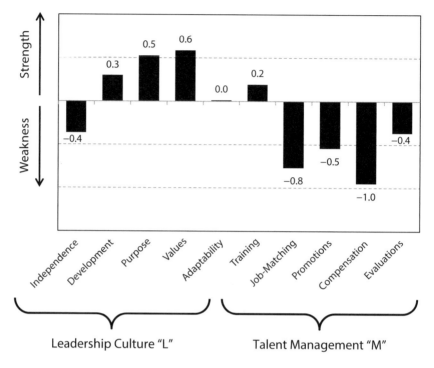

FIGURE 1.1. Leader/Talent Scores for the US Military
Source: Author's Leader/Talent survey

talent management. The strongest categories were values and sense of purpose in the military; the weakest categories were job-matching, promotions, and compensation. For example, one of the lowest average marks by military members was on the statement, "Pay is closely aligned with performance."

One word of caution: any organization can deflect criticism by describing itself as unique, with a unique mission, unique constraints, and so on. This is a reflexive instinct among military leaders—many of whom have not worked for an employer other than their service branch since age twenty-one. Many military leaders transcend this reflex, and here we will provide an analytical instrument for self-examination of an organization's strengths and weaknesses relative to itself and its history.

A savvy reader might wonder why we assume there are only two dimensions of organizational performance. A third dimension might be relevant, perhaps more relevant, to organizational performance. In fact, empirical analysis of responses to the forty elements in my survey split rather neatly into two fundamental areas (not more), a result that drives our categorization.

Defining Organizational Leadership and Management

In this chapter, "leadership" is distinguished from management to include all the elements of organizational culture. "Management," on the other hand, refers to only personnel/talent/human resource (HR) policies. The phrase "talent management" was coined in a 1997 McKinsey study and then a 2001 book by Ed Michaels, Helen Handfield-Jones, and Beth Axelrod called *The War for Talent*. Although that so-called war is often perceived as a recruiting battle, that is only the starting point for talent management, as Charles O'Reilly and Jeffrey Pfeffer articulated in their 2000 book *Hidden Value*. When concerns about low retention rates of some talented military officers and soldiers arose in 2006, the term "talent management" became common in military circles. In the years since, it has become synonymous with personnel policies in general, not just retention policies for top employees.

A reading of top papers on these topics from the *Harvard Business Review,* textbooks on personnel economics, and material on human resources management guided the development of the Leader/Talent matrix. A *well-managed* firm is considered one that has effective personnel policies. The well-managed firm has thought carefully about how to treat employees fairly, with good training programs, ample use of merit-based compensation, and useful performance evaluations. It optimizes firm performance by getting the right people in the right jobs with a level of independence that balances

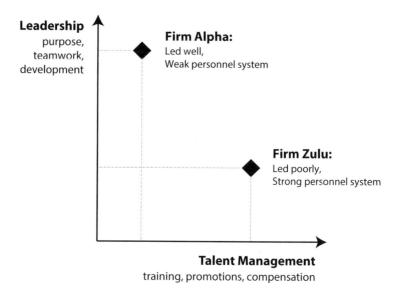

FIGURE 1.2. Dimensions of Organizational Leadership and Management
Source: Author's Leader/Talent survey

creativity and focus. Likewise, effective policies give employees enough control over their own careers so that they can specialize or generalize just enough to maximize morale without distracting from overall productivity.

On the other hand, *well-led* firms develop a strong team ethos in the classic definition of leadership: assembling a group of people in a coordinated effort to achieve a common goal. In business, that usually means producing something of top quality; in the military, that means accomplishing a mission. But in all organizations, leadership encompasses independence, values, common purpose, adaptability, and individual development.

In this view, organizational leadership and organizational management are different but not opposites. An organization can be well (or poorly) managed and well (or poorly) led or any combination. In practice, organizations are more complicated mixtures than the four simple theoretical states, and we can imagine two quantifiable dimensions, shown as axes in figure 1.2. Firm Alpha in the figure has a relatively high measure of leadership but a low measure of talent management policies, much like the US military.

Leader/Talent Survey

The Leader/Talent employee survey was developed with an advisory team of scholars, business executives, and military officers. The survey includes a list of distinct elements that describe the organization's culture and talent practices. The basic structure of an element is a descriptive statement. Respondents are asked to evaluate their employer in terms of each element using a five-point scale (table 1.2). For example, one element is "Young leaders are given serious responsibilities." The five possible responses are "always true" (2), "often true" (1), "sometimes true/neutral" (0), "often false" (−1), or "always false" (−2). The elements are randomized in each survey, and respondents do not know which elements are in which categories.

With input from an advisory team and trial runs with focus groups, the phrasing of elements was refined in 2014. For example, one senior military officer insisted, rightly, that the L/T matrix should try to measure how an organization deals with poor performers. Could they be easily discharged? Were they never promoted? The full, final forty elements in the Leader/Talent survey are described in the appendix. About half (nineteen) of the elements fall under leadership culture, and the remainder (twenty-one) fall under talent management practices. The forty elements comprise five broad leadership categories and five broad talent management categories.

Some aspects of managing people do not neatly fall along one or the other dimension; none was more difficult to describe than feedback from a boss to a subordinate. Although most aspects of leadership

TABLE 1.2.

Element Response Options

2	Always True
1	Often True
0	Neutral
−1	Often False
−2	Always False

Source: Author's Leader/Talent survey

development seem to be cultural, such as mentoring, there are distinct aspects of performance appraisals that are central to talent management and the human resources bureaucracy. In addition to the forty Leader/Talent elements, the survey also asks five questions about organizational performance and the qualities of top officers. A full description of the study, methodology, and advanced statistical analysis is available online.[15]

The Leader/Talent survey was promoted in mid-2014 using an online survey platform that promised anonymity to respondents. The initial contacts had participated in a West Point graduate survey.[16] Class presidents/secretaries of graduating classes from the US Naval Academy and US Air Force Academy were asked to distribute the survey. Participants were encouraged to propagate a web link among fellow veterans of all ranks, which was also made public on social media. In late 2014, two organizations participated in the survey. The first was a profitable technology firm; the other was a class of high-potential senior US Air Force officers. Lastly, on April 1, 2015, the *Military Times* included another public link to the survey in its daily "Early Bird Brief" e-mail newsletter; that link was also posted by the prominent MarginalRevolution.com economics blog.

There are 566 respondents in the final, clean dataset (including 167 who provided assessments of two different employers), yielding 733 total unique observations. Of those, 360 served on uniformed active duty (others worked for the military in a different capacity); 244 respondents evaluated their private-sector employers; and there were numerous observations of government and nonprofits.

Responses to the survey were almost entirely opt-in, with over two-thirds coming in April 2015 through a web link promoted in a military publication's daily newsletter and a same-day reposting on prominent blogs and Twitter accounts. That kind of sample can be vulnerable to self-selection bias. Although bias in any survey is impossible to remove completely, there are ways to test for and manage it.

The biggest concern is that voluntary respondents to a survey are more interested in the subject than nonvolunteers, making the volunteers either too critical (sour grapes) or too enthusiastic. However, that kind of bias will not have any impact on intervariable differences. Specifically, respondents to the Leader/Talent survey consistently show similar relative patterns in rating the forty elements, which are presented in randomized order for each respondent. For example, the element with the highest average score among military respondents, at $+0.84$, was "This organization has a strong sense of purpose." That compares to the lowest, at -0.91: "Bonuses are used effectively to reward good work." Even if one group, such as US Navy officers, gives responses that are upwardly biased in general by a half point, that does not distort which elements are internally perceived as strengths versus weaknesses.

Consider the patterns across the different branches of the military. The Army and Air Force got their highest marks on values $(+0.6, +0.5)$ and lowest on compensation $(-1.0, -1.0)$. Even though the hundred-plus Army respondents were generally more positive about leadership culture than the hundred-plus Air Force respondents, the pattern among the Leader/Talent categories looks like a carbon copy.

Another concern is that veterans who leave the military after their first commitment (around five years) might be more critical than active-duty troops. Yet active-duty respondents in the sample made up three-quarters of military observations and were *more critical* than veterans, consistently across ranks, including veterans who served until retirement (twenty-plus years of service). Thus we see in table 1.3 that active-duty members rated their military employer as weaker in terms of retention ("retaining the most talented people," where a 10 rating is the best) than veterans by a large margin. Enlistees on active duty (with an average of fifteen years of service) gave the lowest average rating; retired senior officers gave the highest rating. Likewise, the average leader element and average management element got much lower average scores from active-duty respondents from all ranks, with the gap ranging from 0.2 to 0.8 lower for ratings by active-duty respondents.

A third check on bias in the survey is two subsamples that validate the overall findings. A rule of thumb among management scholars is that the best way to determine weaknesses in organizational culture is to talk with high potentials: the young executives most likely to be promoted to senior ranks. A group of high-potential Air Force

TABLE 1.3.
Comparison of US Military Respondents

	Enlisted		Jr. - Mid Officer		Sr. Officer	
	Veteran	Active Duty	Veteran	Active Duty	Veteran	Active Duty
N	24	14	28	147	32	85
Yrs. of Svc.	13	15	10	12	20	20
Retain	5.2	3.6	4.6	4.2	6.7	5.6
Avg. "L"	+0.4	−0.4	+0.6	0.0	+0.7	+0.4
Avg. "M"	−0.3	−0.9	−0.4	−0.7	−0.1	−0.3

Source: Author's Leader/Talent survey

officers at a selective service school was asked by the dean to participate anonymously in the Leader/Talent survey (all but one did). Responses were nearly identical: the average L element score was +0.19 for the nonvolunteer group compared to +0.15 for other Air Force respondents; the average M score was 1.93 versus 1.79 for others.

As described in the introduction, the Leader/Talent survey results proved to be highly indicative of actual strengths and weaknesses. The average US military score across all leadership elements was +0.5, whereas the average talent management element was −0.5 using a 4-point scale from +2.0 to −2.0. The weakest military categories are job-matching, promotions, and compensation. The average score for job-matching is −0.8 for the armed forces,[17] which stands in stark contrast to +0.5 for the private sector. Military performance in those three categories is weak, not just relative to the private sector but compared to its own performance in other categories.

Figure 1.3 presents the scores of the three largest US military branches across all ten categories. The highest scores are in the values and purpose categories. In fact, the military's +1.0 score for one of the purpose elements ("This organization has a strong sense of purpose") was the highest score on any element. The largest branches had very similar results with a few notable exceptions. In general, the USAF sample has more negative perceptions, which may or may not be representative of the larger USAF population. What is significant is the extremely low USAF scores relative to the other branches in independence and adaptability (e.g., "The focus here is on mission success, regardless of barriers"). On the other hand, the Navy scored relatively higher in evaluations, most notably, and a handful of other categories.

Assessments were consistent across all ranks—colonels, lieutenant colonels, majors, captains, and enlistees—as shown in figure 1.4. Generally, more senior ranks are more positive for all Leader/Talent categories, but there is broad agreement across the ranks about what is working well versus not working well. Consistency across the

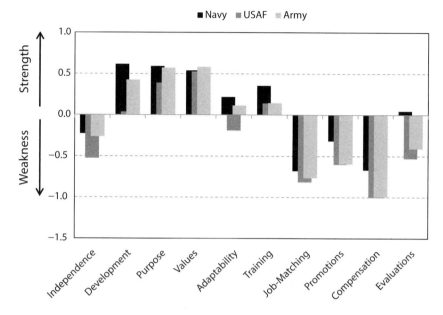

FIGURE 1.3. Leader/Talent Scores by Service
Source: Author's Leader/Talent survey

ranks is perhaps the most striking result in this study; every military rank considers talent management policies, not culture, to be the weak links.

In sum, the Leader/Talent survey provided exactly what it was designed to provide: a target for the areas ripe for reform. Figures 1.5 and 1.6 report detailed scores for each element. One of the strongest signs is a perception among military members that "there are many great leaders in the organization." However, people do not feel they are encouraged to take risks, but instead that conformity is rewarded more than creativity. Other questions suggest that officers are hardworking but not entrepreneurial.

The average score on compensation does not mean that the military paychecks are low. Rather, it means that the compensation process

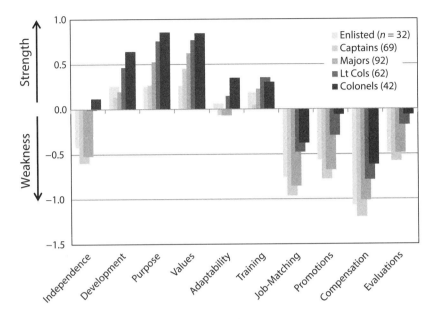

FIGURE 1.4. Leader/Talent Scores by Military Rank
Source: Author's Leader/Talent survey

is evaluated poorly on four elements: rewards to performance, good work, efficiency, and retention. Among those four elements, the military-civilian gap is largest on "Pay is closely aligned with performance" and "Bonuses are used effectively to reward good work."

Across the board, members see promotion practices in a negative light. However, the five elements allow reformers to distinguish which practice is weakest, which in this case is "Poorly performing employees are never promoted." That suggests that active-duty troops are more upset by seeing weak leaders promoted quickly than by seeing great leaders promoted slowly.

The full Leader/Talent study, which is available online, also sought to identify which of the elements actually matter for performance. For example, what correlates with talent retention? It would be a

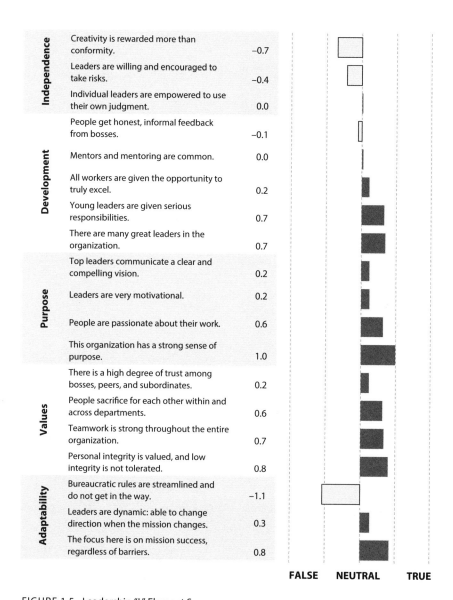

		Score
Independence	Creativity is rewarded more than conformity.	-0.7
	Leaders are willing and encouraged to take risks.	-0.4
	Individual leaders are empowered to use their own judgment.	0.0
Development	People get honest, informal feedback from bosses.	-0.1
	Mentors and mentoring are common.	0.0
	All workers are given the opportunity to truly excel.	0.2
	Young leaders are given serious responsibilities.	0.7
	There are many great leaders in the organization.	0.7
Purpose	Top leaders communicate a clear and compelling vision.	0.2
	Leaders are very motivational.	0.2
	People are passionate about their work.	0.6
	This organization has a strong sense of purpose.	1.0
Values	There is a high degree of trust among bosses, peers, and subordinates.	0.2
	People sacrifice for each other within and across departments.	0.6
	Teamwork is strong throughout the entire organization.	0.7
	Personal integrity is valued, and low integrity is not tolerated.	0.8
Adaptability	Bureaucratic rules are streamlined and do not get in the way.	-1.1
	Leaders are dynamic: able to change direction when the mission changes.	0.3
	The focus here is on mission success, regardless of barriers.	0.8

FALSE   NEUTRAL   TRUE

FIGURE 1.5. Leadership "L" Element Scores
Independence (encouraging individual judgment, risk, and creativity)
Development (early responsibility, opportunity, and mentorship)
Purpose (passion, group purpose, and shared vision)
Values (teamwork, trust, and sacrifice)
Adaptability (mission focus, dynamism, and flexibility)
Source: Author's Leader/Talent survey

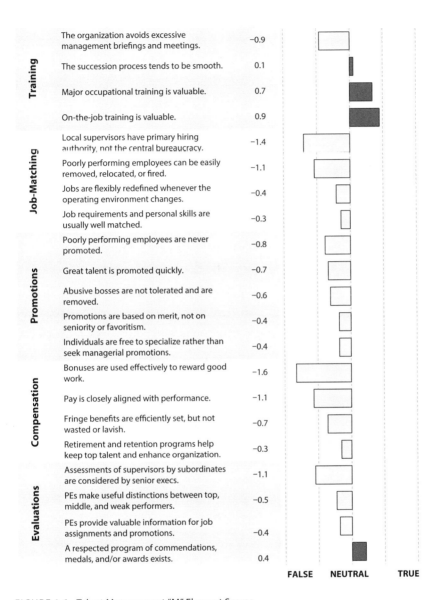

		Score	FALSE	NEUTRAL	TRUE

Training
- The organization avoids excessive management briefings and meetings. −0.9
- The succession process tends to be smooth. 0.1
- Major occupational training is valuable. 0.7
- On-the-job training is valuable. 0.9

Job-Matching
- Local supervisors have primary hiring authority, not the central bureaucracy. −1.4
- Poorly performing employees can be easily removed, relocated, or fired. −1.1
- Jobs are flexibly redefined whenever the operating environment changes. −0.4
- Job requirements and personal skills are usually well matched. −0.3

Promotions
- Poorly performing employees are never promoted. −0.8
- Great talent is promoted quickly. −0.7
- Abusive bosses are not tolerated and are removed. −0.6
- Promotions are based on merit, not on seniority or favoritism. −0.4
- Individuals are free to specialize rather than seek managerial promotions. −0.4

Compensation
- Bonuses are used effectively to reward good work. −1.6
- Pay is closely aligned with performance. −1.1
- Fringe benefits are efficiently set, but not wasted or lavish. −0.7
- Retirement and retention programs help keep top talent and enhance organization. −0.3

Evaluations
- Assessments of supervisors by subordinates are considered by senior execs. −1.1
- PEs make useful distinctions between top, middle, and weak performers. −0.5
- PEs provide valuable information for job assignments and promotions. −0.4
- A respected program of commendations, medals, and/or awards exists. 0.4

FIGURE 1.6. Talent Management "M" Element Scores
Training (occupational, on-the-job, and recurring)
Job-matching (local control, efficiency, and removal)
Promotions (merit, differentiation, and specialization)
Compensation (merit, bonuses, and benefits)
Evaluations (usefulness, peer/360 degree, and recognition)
Source: Author's Leader/Talent survey

waste of resources to fix training programs if that is not actually linked to retention. The question about organizational performance was phrased, "Using any number from 0 to 10 (where 10 is the best in the world) rate this organization at

... retaining the most talented people.

... recruiting excellent people.

... making its employees happy and proud about their work.

... making the best product/service for customers.

... getting the highest quality work possible from the employees it has."

Results showed the military to be strongest at recruiting and weakest at retention on quality work. The key result is what Leader/Talent categories actually matter for performance. The data showed three to be especially important for all aspects of performance: purpose, values, and job-matching. In general, the five performance metrics correlated more significantly with leadership aspects of an organization than with its management aspects. However, the issue of retention was more nuanced. The military's strong sense of purpose is the most significant aspect for retaining talent, followed closely by promotions and compensation. The fact that the latter two are perceived as weaknesses in Pentagon talent management should give reformers a clear agenda for change.

Leader/Talent Results and Conclusion

An analysis of Leader/Talent data confirm the hypothesis that leadership culture is distinct and complementary to talent management. Indeed, organizations that tend to be strong in one category of leadership will also tend to be strong in other categories, and will also tend to score higher on talent management categories as well (figures 1.7 and 1.8).

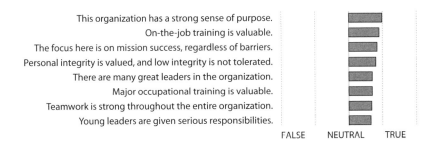

This organization has a strong sense of purpose.
On-the-job training is valuable.
The focus here is on mission success, regardless of barriers.
Personal integrity is valued, and low integrity is not tolerated.
There are many great leaders in the organization.
Major occupational training is valuable.
Teamwork is strong throughout the entire organization.
Young leaders are given serious responsibilities.

FALSE   NEUTRAL   TRUE

FIGURE 1.7. Best Elements from Military Responses
Source: Author's Leader/Talent survey

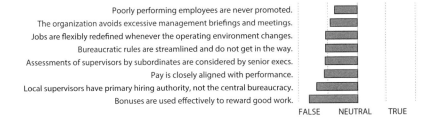

Poorly performing employees are never promoted.
The organization avoids excessive management briefings and meetings.
Jobs are flexibly redefined whenever the operating environment changes.
Bureaucratic rules are streamlined and do not get in the way.
Assessments of supervisors by subordinates are considered by senior execs.
Pay is closely aligned with performance.
Local supervisors have primary hiring authority, not the central bureaucracy.
Bonuses are used effectively to reward good work.

FALSE   NEUTRAL   TRUE

FIGURE 1.8. Worst Elements from Military Responses
Source: Author's Leader/Talent survey

The US armed forces are very similar to one another in their Leader/Talent scores, which is interesting because they are truly distinct organizations, although they operate under the same set of management rules by law and regulation from the federal government.

The military services score far lower on three talent management categories than on any of the leadership categories. This finding is confirmed in a subsample of high-potential USAF officers who were not self-selected. The finding is also confirmed in the trend lines of five military ranks; colonels, lieutenant colonels, majors, captains, and enlistees have different levels of ratings, but their trends across the ten Leader/Talent categories are nearly identical.

At a minimum, the findings here substantiate and can help focus contemporary concerns in the US military about talent management reform. The next chapters provide an overview of current talent management policies followed by recommendations that truly empower troops by creating a total volunteer force.

CHAPTER TWO

Background

The United States military is one of the largest and most complex organizations ever created. It has sustained a level of excellence and dominance unparalleled in history. In order to maintain its superiority, there is a continual pressure to modernize personnel practices just as the military services modernize weaponry, logistics, and strategy. The Pentagon's personnel practices are shaped by ancient cultural traditions as well as major reforms such as the 1973 introduction of the All-Volunteer Force (AVF) and the 1980 Defense Officer Personnel Management Act (DOPMA). With 1.3 million volunteers on active duty, military systems are vast and consequential—and the subject of countless opinions pulling in different directions. Despite their excellent performance during times of war and peace, the armed forces' management of the human dimension is challenged by dramatic budgetary cuts, even as the labor market is shifting in the new global economy.

Many myths about service members proliferate even among defense officials and troops themselves. A widespread concern is that only 1 percent of Americans serve in uniform, cited by many as evidence of the civilian-military cultural gap. While it is true that fewer

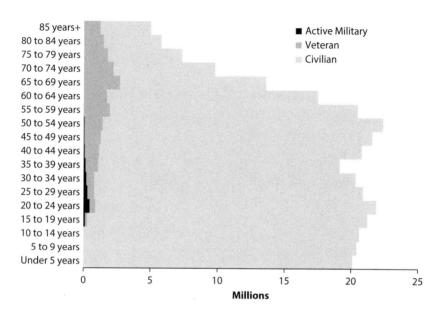

FIGURE 2.1. US Civilian-Military Population Pyramid
Source: Author's calculations based on US Census, ACS, and DMDC (most recent data as of 2015)

than 1 percent of Americans are on active duty—1.3 million from a population of roughly 320 million people, of which 246 million are adults (age eighteen and older)—multiples more are veterans. Three million Americans were on active duty during the peak years in Vietnam[18] and more than 21.3 million veterans live in the United States, according to the US Census Bureau. Once all veterans are included, the number of Americans who are "military" is actually 8.7 percent. And the percentage of military is much higher if one includes immediate families of veterans (i.e., spouses, siblings, and young children).

One out of eleven Americans, not one in a hundred, is or has been on active duty (figure 2.1). What other organization has employed such a high percentage of Americans? One might charge that there is a civilian-firefighters gap because so few citizens are full-time firefighters, and the same logic could apply to police, to doctors, or to engineers. Regardless, some believe this gap makes it easier for the government to engage in war, hinting that a draft would deter wars. Secretary of Defense Ash Carter recently noted that in past generations three in four families had a member in the military, whereas now it is only one in three.[19] The concentration of military families may be a concern, but it may also be a positive sign of professional respect among a unique community of Americans. Similar familial patterns occur for other professions. An essential point about the civ-mil gap is what it is not. Respect for the institution of the military among American citizens is higher than for any other. It actually increased in the decade after creation of the AVF, unlike the degradation of esteem for nearly all other institutions.

Another persistent myth is that the AVF relies on recruiting low-skilled, young men in poor, urban areas—a bogus story that was spread during the early stages of the Iraq war. "Very few" of the soldiers fighting in Iraq "are coming from the privileged economic classes," reported the New York Times on August 18, 2005, echoing similar stories in the Washington Post, the Los Angeles Times, and the controversial Michael Moore film, Fahrenheit 9/11. In fact, the US military's high intellectual and physical standards mean that fewer than one in eight young Americans can even qualify to enlist. High school graduation rates for enlistees are 97 percent, compared to the civilian rate of 80 percent. Moreover, during the heat of the Iraq war, there were three enlistees from the wealthiest US neighborhoods for every two enlistees from the poorest neighborhoods. The reality is that the quality of today's service members is extremely high.

Nevertheless, top government officials routinely repeat misinformation out of ignorance. Senator John Kerry joked in 2006 that American students who neglected their classwork would end up "stuck in Iraq," which echoed his generation's draft-era view of enlistees. In 2015, Ben Bernanke, the former chairman of the Federal Reserve, made a similar claim during a public discussion about the military-industrial base. Bernanke incorrectly paraphrased an academic study when he suggested that military service was arguably a career-skill setback for the typical eighteen-year-old, despite the fact that the study he was referring to was a critique of the *draft* and its effect on people who graduated from high school in the 1960s. The persistently high unemployment rate for younger veterans evaporates quickly after a few years. A major study by the Rand Corporation confirmed that veterans have much lower rates of unemployment than civilian peers after a transition period during which ex-military members learn how civilian markets work.[20] This leads to higher incomes for veterans, confirmed by a recent study from the Department of Veterans Affairs that found "post-9/11 veterans attain 11 percent higher median earnings than non-veterans with similar demographic characteristics," an advantage that was even higher for female veterans.[21]

While considering changes to personnel policies, it would be wise to set aside conventional wisdom about alienation and exploitation. Later chapters will more carefully analyze strengths and weaknesses of those policies and recommend a number of reforms, but this chapter aims to provide essential context. First, we will consider the basic but often misunderstood labor markets in which the military labor force exists. Other sections will summarize changes in the overall size and demographics of military manpower, organizational structures, standard career paths, and military finances.

Changing Labor Market Dynamics

The "battle space" for personnel operations is the competitive labor market, principally the large civilian labor market in the United States. This battle space is as complicated with fog and friction as any other. Even though the armed forces operate as a closed economy in many ways, adopting the AVF in 1973 required each service branch to compete for talent during the recruiting phase, which led to, first, the advent of marketing for military recruits and, second, major changes in military compensation.

The United States has experienced three recessions in the past three decades, in 1991, 2001, and, most recently, in 2009. Each time, the unemployment rate rose by 3–6 percentage points, exceeding 10 percent only once. Despite these recessions, an important underlying development in the US economy is that the mix of jobs has been evolving rapidly. Higher skills are in greater demand while lower-skill jobs have become relatively rare. In short, there is a rising demand for the kinds of skills that are needed in the US military, a mixture of technical aptitude and the soft skills of personal responsibility and teamwork.

There are constant dynamics in American labor markets. For example, the federal government reported that 7.6 million jobs were created in the second quarter of 2015; simultaneously, 6.7 million were lost.[22] Longitudinal studies confirm that Americans tend to change employers many times during their careers. On average, baby boomers worked for a dozen different employers during their early working decades, and only one in ten workers had fewer than five employers.[23] The US military's careerist culture is anachronistic, but also cut off from realities of the modern talent ecosystem.

The Internet transformed open (not just private-sector) labor markets, first with job-matching websites such as monster.com and more recently with professional social networks, of which LinkedIn is the most ubiquitous. Traditionally, jobs were listed in newspapers, and job-matching was a lengthy process dominated by word-of-mouth information. Job candidates were considered using printed résumés, often mailed or hand-delivered to hiring companies. Background checks were conducted with phone calls, not Google searches. Within a decade, the Internet shattered the traditional model and slashed traditional print media revenues in the process. Today, networks still matter, but information-sharing is instantaneous on the web. Some believe the LinkedIn profile will replace the résumé altogether.

Today the competition for talent by private markets makes it even more challenging for the armed forces to keep their labor pools separate, making retention the pressing issue of our time, just as recruiting was the pressing issue of the 1980s. Thanks to the Internet, awareness of private-sector opportunities among active-duty troops is higher than ever. Unlike a soldier deployed to a foreign base in 1995 or 1975, a soldier in 2015 had real-time awareness of job opportunities outside the military.

Military Manpower

In addition to the 1.3 million Americans on active duty, many others serve in the Coast Guard, Air and Army National Guard, the reserves, and as civilians. Reserves (including National Guard) are 811,000 strong and, after a decade at war, are expected to maintain the same standard of readiness as their active-duty counterparts. Civilians serving the armed forces total 773,000 individuals.[24]

Among the services, the Army is the largest branch with 475,000 uniformed service members, and the Navy is second with 327,300. The Air Force is the third largest with 317,000, and the Marine Corps has 182,000.[25] The active-duty force is composed of 1.06 million enlisted members and 229,000 officers (table 2.1).

The 1.3 million active-duty troops in 2016 compare to 2 million in 1990 (the end of the Cold War) and 3 million in 1970 (the middle of the Vietnam War), as shown in figure 2.2. The figure shows that the total number of active-duty service members doubled between 1950 and 1953, peaking at just over 3 million in the early 1950s and early 1970s, and drawing down gradually over the past four decades.

Two surprising facts about the US force posture in recent years stand out. First, the percentage of the US population serving on active duty is lower today than at any time in the modern era, currently

TABLE 2.1.

Military Strength by Service and Rank in FY 2016

Force	Active Total	Officers	Enlisted	Cadets	Reserves	Civilians	Total Force
Army	475,000	91,941	378,581	4,478	198,000	205,416	878,416
Navy	327,300	54,333	268,524	4,443	57,400	180,780	565,480
Marines	182,000	20,912	161,088	n/a	38,900	20,264	241,164
Air Force	317,000	61,690	251,310	4,000	69,200	169,888	556,088
Air and Army National Guard					447,500	n/a	447,500
Pentagon						196,998	196,998
Totals	1,301,300	228,876	1,059,503	12,921	811,000	773,346	2,885,646

Source: Office of Management and Budget, Department of Defense-Military Programs, FY 2017, 223–224

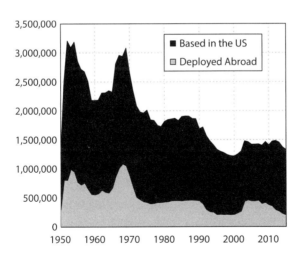

FIGURE 2.2. US Active-Duty Troops, 1950–2015

Source: Defense Manpower Data Center, author's calculations

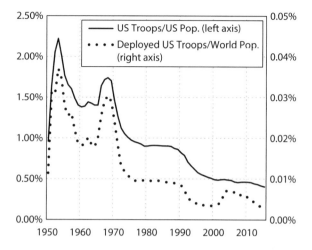

FIGURE 2.3. US Active-Duty Troops Relative to Population, 1950–2015
Source: Defense Manpower Data Center, author's calculations

less than half of 1 percent—0.43 percent to be exact. Second, today there are fewer deployed US troops based overseas relative to the world population than at any time since 1950 (see figure 2.3). These two facts establish an unmistakable long-term trend: the strategic withdrawal of US forces from the world. Regardless of public impressions of heavy US engagement, the downward trend seems beyond doubt.

One reason today's military is smaller compared to the past is the higher standards required to enlist, though this may be more effect than cause of the AVF. The Department of Defense (DOD) estimates that only 13 percent of young US citizens are likely to meet all the military's qualifications without a waiver,[26] beginning with a minimum score on the Armed Forces Qualification Test (AFQT). Enlistees can have no history of drug use or criminal activity and must be US citizens or legal residents. Each service has additional, specific requirements. The Air Force, for instance, can disqualify candidates for having more than 25 percent of their body covered in tattoos.

Army recruits must meet body fat content levels of 20–26 percent for men and 30–36 percent for women.[27] One DOD study found that 22 percent of applicants were turned down for medical reasons, 21 percent because they exceeded the weight limits, 14 percent for mental health issues, and 8 percent for drug use.[28]

The projected size of the future force has been reduced in recent years due to major budgetary restrictions. For example, the Marine Corps maintains a state of readiness with twenty-three infantry battalions, down from twenty-five battalions in fiscal year (FY) 2014. The Marine Corps received additional funding for FY 2016, but troop levels will decrease from 184,000 to 182,000, and possibly to 175,000 if the budget cuts known as sequestration remain in place through 2019. Table 2.2 provides further details on the future shape of the entire force.

TABLE 2.2.
Projected Military Manpower

Service	FY 2015	FY 2019
Active Army	490,000	440,000
Reserve Army	198,000	195,000
Army National Guard	350,200	335,000
Active Navy	323,600	323,200
Reserve Navy	57,300	58,800
Marine Corps	184,100	182,000
Active Air Force	313,000	308,800
Reserve Air Force	67,100	66,500
Air National Guard	105,000	103,600

Source: FY 2015 and FY 2016 President's Budget Requests

Demographics

The major demographic changes in military manpower are the increasing participation of women and the increasing number of families (table 2.3). In 1973, women represented only 2.2 percent of the enlisted force. The number of women in uniform rose dramatically at the end of the century and, since 2000, has remained steady at around 16 percent. Even so, retention rates for women are notoriously lower than for men, which is a vexing problem for the services.[29]

Fifty-five percent of the current active-duty force is married, up from 40 percent in 1973. Among total active-duty members, more than 11 percent are in dual-military marriages; more specifically, 20 percent of all active-duty females are married to another military member.[30] Trying to manage dual-career assignments is a challenge that previous eras essentially did not face. Furthermore, more than a third of active-duty personnel have children.

Many Americans believe that the present military system targets and recruits larger proportions of low-income and minority families. Research, however, suggests that military recruits come from all income levels, with a majority from middle- to high-income neighborhoods. A 2008 study based on 2006–07 data from the DOD's Defense

TABLE 2.3.
Family Demographics in the US Armed Forces

Force	FY 2014 Totals	Female	Married	Dual Marriage	Children	College	Average Age
Army	504,330	13.9%	59.3%	5.3%	48.6%	15%	29.4
Navy	321,599	17.8%	51.6%	5.1%	38.3%	10%	28.7
Air Force	312,453	18.9%	58.3%	11.3%	42.9%	13%	25.4
Marines	187,891	7.6%	45.7%	3.8%	30.5%	10%	29.3

Source: US Department of Defense, Demographics Report 2014

Manpower Data Center found that 11 percent of enlisted recruits came from low-income families, while 25 percent came from high-income families. A 2005 study found that three enlistees come from America's richest neighborhoods for every two who grew up in the poorest neighborhoods. Finally, nearly 40 percent of enrollees in the nation's ROTC programs are from the wealthiest neighborhoods.[31]

Capital Deepening

Warfare is under continuous evolutionary pressure, most visibly in the arms race among nations. The pressure to innovate manifests in a new generation of tanks, ships, and aircraft, but also in the increasing ratio of equipment to personnel. This "capital deepening" not only means that the cost and capability of new hardware are rising, but they are rising relative to manpower.

After the Cold War, there was debate on how to transform Army units in order to allow for greater mobility and flexibility to meet a new variety of threats. By 2005, the Army had begun the transition to Brigade Combat Teams (BCTs), which are smaller and more self-sufficient than traditional brigades. A BCT typically includes three infantry or armored mechanized battalions, a field artillery battalion, an engineer battalion, a support battalion, and a reconnaissance battalion.[32] In 2010, the Army introduced Combat Aviation Brigades (CABs) as another modular unit. CABs have full-spectrum capabilities, which include a headquarters, assault helicopter battalions, support battalions, a reconnaissance squadron, and several companies. The Army plans to decrease end strength from 491,000 to 450,000 by FY 2019, including a reduction in civilian employees and a reduction in the manning of some BCTs by 1,000 soldiers (one quarter of the current size).

The Navy had plans to increase the fleet from 300 ships to 375 ships over a decade ago, but the number of vessels instead continued its long downward trend. As a generation of older surface ships retired, the replacement rate became less than one. Another factor diverting resources was the Pentagon's focus on land wars in Iraq and Afghanistan. Even so, capital deepening is ongoing in the Navy. Restoration of the USS *George Washington* will bring the number of aircraft carriers back up to eleven. Meanwhile, whole new ship classes are being introduced, such as the littoral combat ship and expeditionary fast transport. Navy Secretary Ray Mabus has emphasized that during his tenure the US Navy fleet will once again be above 300 by the end of the decade.[33]

The Air Force's size peaked at roughly 9,000 aircraft in 1989, but has been cut in half in recent years to fewer than 3,900 aircraft. At the same time, manpower has been cut by 40 percent. Air Force wings are traditionally composed of ten squadrons, with three to five of them operational and others support of some kind. Squadrons vary in purpose and function, depending on the type of aircraft, and have undergone some changes since the 1990s. For instance, in 1992 the Air Force reduced the size of squadrons from twenty-four to eighteen aircraft in order to increase their flexibility and deployment readiness. Even with the reduction in squadron size, the number of squadrons decreased throughout the 1990s. For example, F-15s and F-16s made up forty-seven squadrons in 1991, but only thirty-three squadrons just six years later. Secretary of the Air Force Deborah Lee James stated in her congressional testimony that the Air Force is currently the "smallest it's ever been since inception in 1947" and the oldest with most of its aircraft inventory averaging nearly thirty years old.[34]

Since 2013, strategy and budget pressures have combined to reshape organizational structures. Table 2.4 shows military unit levels from 1989 to 2016, essentially the last year of the Cold War to the present. Unit levels gyrated from 1989 highs (fifty-four Army brigades, more than 8,200 USAF aircraft, 592 Navy ships) to peace dividend lows in 2000 (thirty-three brigades, 4,900 aircraft, 316 ships) to

TABLE 2.4.
Active Force Levels

	FY 1989	FY 1995	FY 2000	FY 2005	FY 2010	FY 2015	FY 2016
Army							
Corps	5	4	4	4	4	3	3
Divisions	18	12	10	10	11	9	10
Brigades	54	36	33	33	48	32	30
Combat Aviation Brigades					13	11	11
End Strength	777,000	508,559	480,000	482,400	547,000	490,000	475,000
Navy							
All Active Ships	592	392	316	282	288	271	272
Aircraft Carriers	16	11	12	12	11	10	10
Navy Surface Warships	212	128	128	111	123	99	99
End Strength	592,000	429,630	367,371	357,853	323,139	323,600	327,300
Marine Corps							
Marine Expeditionary Forces	3	3	3	3	3	3	3
End Strength	196,000	174,561	172,955	179,836	202,612	184,100	182,000
Air Force							
Aircraft Inventory	8,244	5,814	4,944	4,764	4,460	3,895	4056
Fighters	2827	1,763	1,595	1,568	1,493	1,887	1,914
Bombers	411	183	181	172	154	154	154
ISR	554	218	185	174	372	436	438
End Strength	570,000	396,382	351,326	349,362	329,640	312,980	317,000
Total End Strength (in thousands)	1,230	1,505	1,370	1,373	1,417	1,310	1,301

Source: President's Budget Requests; Arsenal of Air Power; Naval History and Heritage Command

wartime surges in 2010 (forty-eight, 4,500, 288) to recent wartime drawdowns mid-sequester in 2015 (thirty-two, 3,900, 271).[35] To be sure, the smaller size of these forces is real, only partly balanced by capital deepening and technological improvements.

This table may reinforce a misguided view of military power. Indeed, the whole notion of capital deepening is often misunderstood in crude terms as an increase in machinery relative to muscle. But industrial descriptions of military force and organizations miss this point that intangible capital is the vital force multiplier of the new century. What should be emphasized is the quality of information technology. Bytes/day processing capability is arguably just as important as sorties/day or throw weights. Perhaps the numbers of wings and brigades are a distraction to some degree.

Skills Planning in the New Century

Skills planning is inevitably an oxymoron over the long term. Central planning of human resources begins with each service projecting the skills (requirements) necessary in the present and future. Once requirements are finalized, then personnel are allocated to training and assignments. This kind of central planning is difficult in a stable operating environment, although basic microeconomic theory shows that centrally planned markets are never superior to open markets. And when the environment is changing, such as the ongoing revolution in military affairs, central planning is next to impossible. Defense experts routinely observe shortages in a third or more of enlisted occupations and overstaffing in another third.[36]

Cyber threats are one of the top national security challenges, according to the 2015 National Security Strategy. Yet ten years ago, cyber threats were not even mentioned in the National Security Strategy or Quadrennial Defense Review.[37] To face the new threats, the Department of Defense has requested $6.7 billion in cyber defense funding for FY 2017. The plans call for the continued construction of the Joint Operations Center for US Cyber Command (USCYBERCOM)— begun just over two years ago—and the creation of 133 Cyber Mission Force (CMF) teams by FY 2018.[38] The objectives of this new force are to defend DOD networks, prevent cyber attacks against the United States, and create full-spectrum cyber options to support military operations and even attack if necessary.[39]

CMF teams are composed of from forty to sixty active and reserve military members, along with civilian and contract personnel.[40] In 2015, the DOD planned to have nearly 6,100 members working in cyber defense by 2018—a goal that has proved challenging due partly to the competitive market for cyber skills in the private sector. As of 2015, it has been reported that only half the cyber operators have been hired, yielding fifty-eight teams, less than half of the goal of 133.[41]

Rapid technological changes are likewise affecting air operations. According to the president's budget request of FY 2017, "readiness decreased in FY 2015 to historically low levels. The continued pressure of deployments and chronic shortage of airmen in critical skill positions are limiting recovery efforts."[42] For example, the Air Force faces a shortage of skilled maintenance personnel, notably in the more advanced platforms.

Unmanned aerial vehicles (UAVs, commonly referred to as drones or remotely piloted aircraft) are another novel military development that involves many unexpected occupational requirements. The UAV industry largely owes its proliferation to DOD's acquisition and development, which has two implications: first, there is no existing body of expertise; and second, demand for that expertise in the private sector is growing rapidly. That means there is pressure on the supply of UAV talent. Currently, the Air Force and Navy make up more than 70 percent of the market for UAVs manufactured within the United States.[43] During 2011, the drawdown of operations in Iraq relieved some pressure on the demand for UAV operations, but the recent surge of the Islamic State and other radical jihadi groups worldwide renewed it. Active UAV[44] assets in the military include:

- 150 MQ-1B Predators—armed, medium altitude, multisensor, primarily used for intelligence
- 93 MQ-9 Reapers—medium altitude, multisensor, ability to conduct precision-guided weapons systems
- 33 RQ-4 Global Hawks—high altitude, near-real-time coverage providing imagery, signals intelligence, and moving target sensors

The Air Force budgeted for a daily maximum of seventy-six combat air patrols (CAPs) in 2016.[45] A CAP is twenty-four-hour persistent coverage on an area of battle space; it may require three or four airframes to cover that CAP, but it could be fewer depending on platforms and mission.[46] The DOD recently announced its plan for a 50 percent defense-wide increase of daily CAP capacity by 2019.[47]

In spite of the size of the UAV fleet, there is a persistent shortage of drone pilots, which is an unexpected and ongoing challenge for the Air Force. According to Colonel Bradley Hoagland, drone pilot positions are 82 percent filled, compared to 100 percent of traditional pilot slots.[48] Hoagland suggests this shortfall is partially a consequence of lower promotion rates to major, not just the nature of operations.[49]

There are 1,300 UAV pilots in the Air Force, a number that is expected to reach 1,650 by 2017.[50] In order to meet that goal, the Air Force announced the creation of a new career track for remotely piloted aircraft (RPA) and increased incentive pay for both officers and enlisted members. Enlisted members would be trained to fly the Global Hawk surveillance aircraft. In time, positions could open for weaponized UAVs.[51] The hope is that in 2017, nearly 400 airmen will graduate from RPA training, double the 192 graduates in the FY 2016 budget.[52] The increase is necessary to offset the high attrition rate of

RPA pilots, which is three times higher than the rate of traditional pilots.[53] The USAF loses about 240 RPA pilots per year, due to long hours, stress, and the reputation as a dead-end job.[54]

And yet, the biggest manpower challenge in the UAV enterprise is the processing, exploitation, and dissemination (PED) of collected information. The Defense Department cannot digest the mass of data (terabytes of streaming video) that it gets from the current number of combat air patrols, a challenge that will worsen when CAPs rise.

A third area that exemplifies the manpower specialization puzzle involves how land forces strategically fight insurgencies. The Army and Marines have gained invaluable knowledge about, and experience with, counterinsurgency techniques over the past ten years.

Today, the Defense Department has 660 Special Operational Force (SOF) teams, three Army Ranger battalions, and a total of 66,000 personnel assigned to Special Operations Command (USSOCOM).[55] The total has risen by 3,000 individuals since 2012 but is expected to be at 70,000 by 2019, according to the 2014 Quadrennial Defense Review.[56] Uncertainty over budgets only amplifies the talent projection challenge. The Army Rangers are acquiring fifty-four more Stryker armored personnel vehicles, but there is not enough logistics support in the SOF ranks to maintain them, according to Roger Carstens, a retired Army Special Forces lieutenant colonel and a senior fellow at Foreign Policy Research Institute.[57]

Naval Special Warfare Command is composed of 10,000 active-duty, reserve, and civilian personnel (up from 8,900 in 2012). This includes special warfare operators, better known as Navy SEALs, special warfare combatant-craft crewmen, and those in supportive roles. Currently, there are ten SEAL teams, two SEAL delivery vehicles, and three special boat teams.[58] The Marine Special Operations Command (MARSOC) has about 3,000 personnel (up from 2,600 in 2012), which consists of 625 critical skills operators, thirty-two teams, and nine companies.[59]

However, the ability to sustain large-scale counterinsurgency and stability operations is limited,[60] as can be seen in the uneven treatment of personnel in programs such as Afghanistan Pakistan Hands (AFPAK or APH), the recent Commander's Emergency Response Program, and the nascent Asia-Pacific Hands Programs. AFPAK started in late 2009 and has been praised for its success and future potential.[61] It provides intense regional language skills, cultural awareness, and counterinsurgency (COIN) training. The program started with thirty-three individuals in 2010, reached 180 by 2012, and totaled upward of 700 members by 2014.[62] The plan for AFPAK Hands was to ultimately consist of 351 soldiers, 144 airmen, 54 Marines, and 117 sailors, divided among seven groups (thirty people in each) for deployment.[63] Individuals were enticed to join AFPAK in recent years with promises that it was a vital new career specialty for the current and future missions. Instead, officers who volunteered for AFPAK subsequently suffered relatively low promotion rates and relatively poor follow-on assignment opportunities.

Laws and Regulations

Federal laws have a major impact on how military personnel management can be run, constraining the services to look almost identical in practice. When legislation relevant to military personnel policy is signed into law, two relevant titles of the US Code are revised: Title 10 (Armed Forces) and Title 37 (Pay and Allowances of the Uniformed Services). This section highlights a few of the laws and regulations that are the most important.

Title 10, subtitle A, Part II deals with personnel and comprises thirty-eight chapters, further divided into hundreds of sections. The number of rules is vast and the complexity is profound. As an example, chapter 43, Rank and Command, includes sections 741 through 750. It is in section 741 where military grades are defined by law, from second lieutenant/ensign up to general/admiral. Any effort to simplify or diversify the grade structure would have to amend this section of the code.

Title 10, chapter 32, establishes maximum numbers of officers allowed to serve in each grade. In fact, it sets different limits for a wide range of total officers. If the number of Army officers totals 20,000, this chapter sets a maximum of 1,613 colonels. If the total is 100,000 officers, then 4,548 can be colonels. And so on. Each service's grade structure is inflexibly defined across dozens of force-size scenarios.

Promotions are subject to guidelines set forth in Title 10, chapter 36. Again, this matters because we must look to the code to assess whether the services are afforded the legal authority to reform their promotion systems—to have any flexibility. It may be that the law already allows flexibility. Unfortunately, much of the code is stifling. Subchapter I enshrines the use of selection boards as the means of promotion for every grade. Moreover, it explicitly limits what information the board is permitted to consider for each officer up for promotion, which "shall apply uniformly among the military departments" (meaning that there can be no exception if a service wants to try something different). Per the text, "No information concerning a particular eligible officer may be furnished to a selection board except for . . . the officer's official military personnel file" and a "written communication" from an officer under consideration. The inclusion of "adverse" information seems to be allowed only if substantiated by an "officially documented investigation or inquiry." Such limitations on information flow are defensible in theory, but indefensible in their universality, with a basic hostility to unorthodox talent in a dynamic world.

Section 616 governs "recommendations for promotion by selection boards" in a manner that defines inflexibility. It explains why the Navy is not permitted to promote two superstar captains to admiral below their "zone" of promotion. Not only do the rigid timetables for promotion limit nuance within the distribution of officers, but each service has the same specific limit on how many officers can be promoted below the zone: 10 percent. The one caveat is that the secretary of defense "may authorize a greater number, not to exceed 15 percent." This is an extreme degree of micromanagement, not only of routine board members, but of the discretion of flag officers, service chiefs, and the secretary of defense.

Even so, only the Air Force maximizes the use of below-the-zone (BTZ) promotions. Some believe that its use creates a halo effect and consequently penalizes late bloomers, while others argue that the lack of BTZ opportunities in other services is discouraging to many of their high-potential officers.

Subchapter II (of Title 10, chapter 36) establishes many more rules on promotions, including rules on eligibility, particularly time-in-grade requirements, competitive categories, and promotion zones. Importantly, this process requires the service secretaries to make long-term estimates of future manpower needs at the microlevel of grade and skill. This is antithetical to a dynamic process. It explains why all the services had difficulty adapting to cyber threats.

Subchapter III establishes rules that discharge officers who fail to be selected for promotion. If an individual is "failed of selection" twice, retirement is involuntary. In short, "up-or-out" is not optional, and there can be no variation of the two-strikes process under the law. While the theory behind the up-or-out principle makes sense for some junior ranks, it is implemented by the services in the extreme. It is not hard to imagine that the proverbial best pilot in the Air Force is forced up and out of the cockpit year after year, while the best acquisitions officer in the Navy is forcibly retired because being a wizard with accounting and contract incentives that save billions of taxpayer dollars is irrelevant if he doesn't get selected for command. In light of how private firms and even government agencies manage their talent, it is hard to justify winnowing talented operational specialists in the armed forces who are not tracked (or interested) in senior management roles.

How DOPMA Hurts Productivity

Many of the legal constraints governing military personnel were instituted following the passage of the Defense Officer Personnel Management Act (DOPMA) in 1980. In concert, its reforms standardized careers across the services and had the effect of institutionalizing a relatively short "full" career of twenty years. The Rand Corporation created a useful online reference at dopma-ropma.rand.org.

Thirty-six years after its passage, the law has many critics. In a critical 2010 report, the Defense Science Board highlighted DOPMA's inflexibility and blamed it for "wasting human capital." A Rand study in 2006 claimed unequivocally that DOPMA-based practices "will not meet the needs of the future operating environment" and called it a "cold war-era personnel system" that was outdated.[64] To be sure, DOPMA was preceded by other laws with far-reaching consequences on personnel flexibility. A few of those are highlighted below.

Before and during World War II, the US Army and Navy had distinctly different personnel systems. Each had far more flexibility to shape its forces on the eve of war as well as mid-war. The Army's chief of staff, General George C. Marshall, famously purged the senior ranks in 1940. He also tapped the relatively low-ranked Dwight Eisenhower—an administrative staff officer in the Pacific—and leapfrogged him over many others to become the supreme allied commander in Europe. Neither action is *permissible* today.

During the war, Army promotions up to lieutenant colonel were decentralized. Field commanders were in charge. After the war, Congress moved to codify the "younger" force that Marshall had achieved. This aversion to old men plugging the upper ranks can be rationalized by asserting that younger minds are more innovative and willing to embrace change, a claim that is scientifically dubious. The Kauffman Foundation's research on entrepreneurial demographics finds that the ages forty to forty-nine are the peak years for starting a new company in the United States. In fact, there are twice as many entrepreneurs in their forties as there are in their twenties. Regardless, the codification of a twenty-year career in the form of the military pension that vests fully at two decades means that a majority of career officers retire around the age of forty-one. That may have been middle-aged in 1945, but it is positively young in our era, and it means the Pentagon is now purging officers with the most entrepreneurial potential.

The Officer Personnel Act (OPA) became law in 1947, extending the "up-or-out" system across the military branches. OPA also made officer promotion boards mandatory and established a universal year-group cohort management process. Finally, OPA initiated voluntary retirement at twenty years of commissioned service.

Manpower patterns changed in ways the Congress had not anticipated, so it passed another law seven years later: the Officer Grade Limitation Act of 1954 (OGLA). OGLA established grade tables for the armed forces, which limited the percentage of officers who could serve in the rank of major and above.

The Selective Service program, which administers conscription in the United States, was established in 1940, disbanded in 1947, then reestablished with the Selective Service Act of 1948. All men are required to register for the draft, or justify an exemption from it, at the age of eighteen. The draft was activated during World War II (1941–45), the Korean War (1950–53), and many years of the Vietnam War (1963–73). President Nixon approved the use of a draft lottery for the first time in December 1969. In 1971, Nixon essentially ended the draft by asking for a two-year extension of the expiring law's authority, so that the last American was drafted in 1973. Many service chiefs resisted the adoption of an all-volunteer force, but it was implemented and became a success after 1973. Two years later, President Gerald Ford took executive action terminating draft registration as well, but his successor, President Jimmy Carter, brought back Selective Service in 1980. It remains in force today.

The Goldwater-Nichols Act of 1986 was the last major piece of legislation to reform military personnel practices. The act shook up the operational command chain, taking the service chiefs out of the direct operational command. Its primary effect on personnel was a requirement that officers could not be promoted to senior ranks without a minimum of one joint duty assignment (e.g., an Army major serving in a job that involves coordination with Navy, Air Force, and/or Marine units) of two to three years in length. The requirement is strict, but bureaucratic definitions of what assignments count often matter more than actual interservice experience.

Personnel Expenditures

The defense budget in FY 2016 is $607 billion, including $145 billion in personnel expenditures and $58.7 billion for overseas contingency operations (OCO) in Iraq and Afghanistan. Under sequestration, defense spending was to be capped at $523 billion in FY 2016, but OCO funding raised it significantly, a process that has been repeated each year.[65] Although about one in four defense dollars (24 percent) are spent on personnel,[66] the defense budget does *not* include billions spent on veterans' care. Funds appropriated for the Department of Veterans Affairs in FY 2016 totaled $167.3 billion, which means the overall spending on military personnel is more than $300 billion.

If DOD and VA spending are added together, four in ten military dollars (40 percent) are spent on personnel. Figure 2.4 displays the total defense, personnel, and VA expenditures in nominal (non-inflation-adjusted) dollars from FY 1980 to 2017.[67] The figure shows

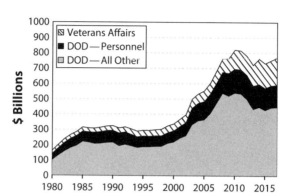

FIGURE 2.4. DOD and VA expenditures, nominal $ (2016 and 2017 estimated)
Source: Office of Management and Budget, historical tables

that nominal military spending peaked in 2010 and has declined during the decade, a decline even starker in real dollar terms.

The outlook for future defense expenditures is cloudy. Although military spending by the United States alone is the equal of the next eight nations combined,[68] it is also true that US spending on defense *as a percentage of GDP* has been cut dramatically over the past few decades. It was above 10 percent of GDP in the 1950s and 1960s but is currently just 4.2 percent (including OCO). Large budget deficits driven by domestic spending have increased US national debt to $19 trillion as of February 2016, a burden that will squeeze defense dollars for the foreseeable future. Many defense officials have described the national debt as the number one national security risk; official projections anticipate annual deficits to rise over the next decade.

Since the Budget Control Act (BCA) became law in 2011, DOD has worked in an environment of fiscal uncertainty. In addition to establishing a Joint Committee on Deficit Reduction, the bill authorized an automatic budget sequestration that set hard caps on spending, most heavily on defense spending in the absence of an alternative arrangement, which the committee was unable to negotiate. Despite

President Obama's promises, the failure to achieve a bipartisan agreement automatically put sequester caps into action.

The effect of sequestration was to slash overall defense expenditures by more than $100 billion between 2011 and 2013. The spending caps set by the BCA have been increased three times, first in 2012, then again in 2013, and most recently by the Bipartisan Budget Act of 2015. In exchange for annual increases, the sequestration caps have been extended until after 2021.[69] The sequestration has affected nearly every aspect of the defense budget, from operations and maintenance to family programs.

Although the Office of Management and Budget (OMB) lists the defense personnel budget as $145 billion,[70] including each branch of service, Guard, reserves, and the Pentagon, we will refer to the slightly different numbers reported in the DOD's M-1 budget, which includes all military personnel programs. The M-1 total enacted budget for FY 2016 is $138.5 billion. This includes line items for officer pay and allowances, enlisted pay and allowances, subsistence, permanent change-of-station (PCS) travel, and dozens of other details for all military members.

Active-duty pay and benefits total $33 billion for officers and $74 billion for enlistees, about half of each being "basic pay." The rest includes retirement accruals, housing, subsistence (meals), social security taxes, and roughly 5 percent for special pay and allowances. Table 2.5 details a few of these items for each service.

In addition to these direct personnel costs, there are other indirect expenditures. For example, the operational budget (O-1) includes $4.5 billion for training and recruiting in the Army, which alone is larger than the entire DOD budget for incentive, special, and allowance pays. A cursory review of the total enacted FY 2016 O-1 budget includes nearly $50 billion in expenditures on human resources, listed in table 2.6.

Of particular interest is the $684 million spent DOD-wide on human resources activity. The cost of managing Human Resources

TABLE 2.5.
DOD Personnel Spending for FY 2016 (in $ millions)

FY 2016	Army	Navy	Marines	Air Force	Grand Total	Total (%)
Pay, Officers	13,499	7,996	2,805	9,075	**33,374**	29.5
...Incentive Pays					**488**	0.4
...Special Pays					**1,189**	1.1
...Allowances					**555**	0.5
Pay, Enlisted	25,046	17,904	8,812	16,893	**68,655**	60.7
...Incentive Pays					**245**	0.2
...Special Pays					**1,730**	1.5
...Allowances					**2,356**	2.1
PCS moves	1,835	923	468	1,235	**4,460**	3.9
Total					**113,054**	

Source: Department of Defense FY 2017 President's Budget (M-1). Grand total does not include reserve or National Guard.

Command is presumably included in the $277 million in the Army category of manpower management and $475 million for other personnel support. The point is that the management of personnel is expensive. The budgetary impact of reforms could easily lower those management costs or raise them, but reforms that decentralize personnel decision-making seem likely to realize large savings in these administrative line items.

Training costs are of particular interest because of the high pace of job rotations in the military. When Dick Cheney was secretary of defense in the late 1980s, he warned that the military was paying a heavy price for the system of frequent job rotations. It is common for a veteran to have served in a number of different units that exceeds the number of years in uniform. Any reforms that slow the rotation pace will harvest savings from lower cross-training costs. In the Army,

TABLE 2.6.
Human Resource Expenditures Not in "Personnel"

	Total (in $ millions)
Defense Human Resources Activity	684
Defense Personnel Accounting Agency	115
DOD Education Activity	2,809
Office of Secretary of Defense (Admin. Services)	1,512
Navy—Training and Recruiting	1,864
Navy—Military Manpower and Personnel Management	347
Navy—Civilian Manpower and Personnel Management	121
Navy—Other Personnel Support	266
Air Force—Training and Recruiting	3,400
Army—Training and Recruiting	4,549
Army—Manpower Management	277
Army—Other Personnel Support	475
Defense Health Program	32,546

Source: Department of Defense, Operation and Maintenance Programs (O-1), FY 2017

recruiting costs are just under 25 percent of the Army's $4.5 billion training budget. Another 15 percent goes to accessions training for new recruits and officers. The bulk, 60 percent, is more advanced training, with nearly $1 billion spent on specialized skills training, $221 million on professional development education, and $570 million on training support. The total annual price of incessant travel and cross-training across all the services might exceed $4 billion; the cost to unit readiness is incalculable.

The Army spends about $500 million to $600 million on recruiting and advertising, which includes radio, television, direct mail, and recruiting pamphlets.[71] Another $175 million is spent on the category "examining," which covers the high school Armed Services Vocational

Aptitude Battery, or ASVAB test, and the US Military Entrance Processing Command (MEPCOM), which measures each potential recruit's medical and moral qualifications. MEPCOM operates more than 450 military testing sites. It is hard to imagine these costs being cut unless one imagines a higher retention rate. If, for example, the average enlisted tenure increased by a year, the military would not need as many new recruits and could also be more selective. The average age of enlistees in 1973 was twenty-five. It rose steadily until peaking in the mid-1990s at around twenty-seven and a half.[72] The average age of officers also rose by two and a half years. It seems likely that a military personnel system that utilized more volunteerism would experience less turnover and therefore lower training costs.

Although indirect costs are higher, the direct costs of rotation are counted in the M-1 budget under the "permanent change of station" category, which was more than $4.4 billion in FY 2016, roughly

TABLE 2.7.
Active-Duty PCS Expenditures, FY 2016 (in $ millions)

PCS Costs by Type	Total Costs
Accession Travel	410
Training	332
Operational	1,124
Rotational	1,730
Separation	669
Travel of Organized Units	45
Nontemporary Storage	53
Temporary Lodging Expense	96
Total	$4,460

Source: Department of Defense, Military Personnel Programs (M-1), FY 2017

4 percent of the military personnel budget. A PCS is the cost of moving a family from one duty station to another if the member will be stationed for six months or longer (table 2.7). Moves shorter than six months are known as temporary duty assignments (TDY). Costs for each PCS include per diems or reimbursements for travel expenditures, temporary lodging expenses, and assistance in real estate purchases.[73] Permanent change-of-station travel also includes separate job-training costs.[74]

CHAPTER THREE

Total Volunteer Force Reforms

This book recommends a series of Pentagon personnel reforms called the Total Volunteer Force (TVF)—an evolutionary step in the same direction as the All-Volunteer Force (AVF) reform of 1973. The heart of the TVF is a *restoration* of authority over personnel decisions to service chiefs and local commanders. In the status quo, authority is extremely centralized in personnel commands such as the Navy Personnel Command based in Millington, Tennessee. It might surprise some outsiders, but many of the innovative ideas to decentralize this authority come from human resources professionals based at these commands.

Many senior officers in the military who have never worked at one of the personnel commands do not acknowledge a problem exists and instinctively distrust reform ideas they perceive to come from outside the military. What must be understood is that extreme HR centralization was imposed from outside the military starting in the 1960s by Robert McNamara, who was appointed as secretary of defense from his role as president of Ford Motor Company. He imposed a centralized personnel system that treated military labor as a commodity rather than a profession. The AVF changed the nature of recruiting after

1973, but the McNamara centralization remained in place. TVF reforms will rebalance personnel authority, but not with a pendulum swing to pre-WWII localization in which field commanders built fiefdoms.

Recommendations

Twenty specific recommendations are presented here that will strengthen the principles of volunteerism and service at the expense of coercion and inflexible bureaucracy. Each TVF recommendation is independent, meaning that one could be successfully implemented alone and improve readiness in the human dimension. They cover three main areas—job-matching, compensation, and performance evaluation—and include evolutionary and revolutionary ideas. In nearly every case, the recommendations will give more flexibility to every service so that, for example, the Marines can hold firm to the personnel structure in place right now, while the Army can transform across multiple dimensions of talent management. Rather than intensify the type of regulations put in place following WWII, the Total Volunteer Force would decentralize personnel management in order to restore command authority.

Roughly half of the recommendations could be implemented with current legal authority, but half require legislative action. Indeed, the most fundamental flexibilities needed to get the right talent to the right jobs can already be implemented: restoring command authority over hiring, implementing web-based job-matching, and extending tour lengths. Likewise, only a few involve monetary costs, whereas the bulk of the reforms have zero fiscal costs and will likely be net-positive in terms of budget impact. Longer careers with higher morale and productivity will yield a stronger military with a lower initial recruitment demand. That will save money but also enable the services to be increasingly selective.

TABLE 3.1.
Personnel Reform Recommendations Matrix

	Evolutionary	Revolutionary
Assignments (Chapter 5)	Restore Command Authority for Hiring & Dismissals Give More Career Control to Individuals / Increase Tour Lengths Implement Web-Based Talent Management / Expand Information Transparency	Restore Service Chief Authority over Promotion Timetables / Flexible Rank Tenure Allow Veterans & Reservists to Apply for Active-Duty Roles Allow Flexible Sabbaticals
Performance Evaluation (Chapter 4)	Use Flexible Ranking in Performance Evaluations	Implement Simple, Multisource (Peer) 360-Degree Evaluations
Compensation (Chapter 6)	Expand Pay Flexibility Retirement Benefit Payments Should Be Later & Larger Training & Education Flexibility	Transform Base Pay from Tenure to Role & Responsibility Expand Retirement Options

Source: Author

The first twelve recommendations focus on various aspects of job-matching, including promotions, permeability, and the end of draft registration (table 3.1). A central tenet of improved talent management is to give more career control to individuals, but the unproductive cultural norms embedded in the current personnel system—such as relentless permanent chang-of-station (PCS) moves and careerism—will not change without legislation to amend the rigid "up or out" timeline set in 1980. Two recommendations offer solutions to performance evaluations, a problem that plagues organizations of all types. The next five recommendations (15–19) touch on compensation reform. Perhaps the most important recommendation is the last one, a plea to the Defense Department and services to conduct a thorough, honest, measurable, regular review of personnel practices.

1. Restore Service Chief Authority over Promotion Timetables

The "up-or-out" principle is so rigid according to the 1980 Defense Officer Personnel Management Act (DOPMA) that every branch of the armed forces promotes officers on the exact same timeline for a decade or more. This law should be revised to allow service flexibility so that the chief of staff of the Army, commandant of the Marine Corps, chief of Naval Operations, and chief of staff of the Air Force can establish promotion rules that are best for their men and women. Even if the Army prefers to maintain the rigid timeline, the Navy (for example) would be allowed to loosen its up-or-out timeline, while the Air Force would be able to end the use of year-group promotion zones entirely.

In general, promotion zones hinder the optimization of job-matching and specialization. Furthermore, if mandatory timelines remain in place, then other reforms will be impeded. However, one mandate should not replace another: each service should be allowed the flexibility to continue using strict cohort promotion zones. If Congress does not amend DOPMA's mandatory up-or-out timelines, it should at a minimum loosen the rigidity of the promotion zones by offering service chiefs flexibility on the issue. Each service should have expansive authority to use below-the zone promotions for up to 40 percent of its officers in each cohort (double the current range).

Legislative action by the Congress is necessary to authorize this.

2. Restore Command Authority for Hiring

Any commander at the rank of O-5 and above should be given final authority on who serves in his or her unit. Personnel centers/commands will provide a slate of no fewer than three candidates for the unit to interview and choose for key roles. Commanders should have limited authority to directly hire, whereas most hires will be through the centrally provided slate of candidates. Many key developmental roles should still be directly assigned centrally—meaning that a single candidate shall be recommended by personnel centers in many instances (e.g., honoring follow-on assignment commitments)—but the unit commander should retain the right to veto a limited number of such assignments.

Each service currently has authority to do this.

3. Restore Command Authority for Dismissals

Allow faster dismissals and quick replacement. Currently, the only way most commanders can remove an individual from the unit is through disciplinary proceedings, and even then they must engage in a lengthy, punitive, and bureaucratic process. The armed forces should empower commanders with a flexible array of options, distinguishing between disciplinary actions and unit fit. The critical missing piece is simply to allow a dismissal for fitness—an action that would not reflect negatively on the service member and would involve a balanced, but less bureaucratic, process. The commander should have discretion to remove (or simply shorten the tenure of) any individual in his unit on the basis of personnel fit, which is distinct from performance and would not circumvent the Uniform Code of Military Justice. Moreover, dismissals and firings of individuals currently leave a hole in the unit that remains unfilled, which penalizes commanders. The dismissal process must be fixed to allow quick replacement.

Each service currently has authority to do this.

4. Give More Career Control to Individuals

Allow individual service members with more than five years of service to:

(1) Opt out of the promotion cycle in order to specialize in their current roles or to apply for roles at their current rank. This would allow, for example, aviators to stay in the cockpit, cyber warriors to remain in critically understaffed positions, and combat commanders to extend their tours during wartime operations in which continuity is vital for mission success.

(2) Turn down the first assignment in a slating cycle without prejudice and with no mark on their record.

(3) Apply for any assignment for which they are qualified. One avenue that should be encouraged is to allow individuals to query unit commanders and HR officers about open and forthcoming jobs.

Each service currently has authority to do this.

5. Implement Web-Based Talent Management

Efficient job-matching requires an information system that makes available jobs (requirements) visible with details about location, unit, role, commander, and more. Likewise, gaining commanders need deeper information about available individuals to optimize matching of faces to spaces. Both sides of the equation need an easy-to-use, unclassified online system, as demonstrated in the Army's "Green Pages" pilot project.

Each service should adopt information technologies that enable troops to enhance their profiles and signal preferences for jobs. The online systems must allow both sides to add supplemental information (i.e., simply allowing individuals to describe additional skills and licenses they have that are not part of existing military records). And the systems must incentivize participation by both sides; a lack of incentives for gaining units will leave most roles ill-defined. These technologies are highly developed in the form of existing, free websites such as LinkedIn and guru.com (among many others).

Optimal job-matching (slating/detailing/assignments) can be decentralized in multiple ways without being completely localized. The status quo for job-matching is generally conducted in discrete periods (usually two or three times) each year, which should be converted to a continuous process. Decentralization allows a dynamic interaction between the supply and demand for troops. Therefore, military personnel commands (e.g., Air Force Personnel Center, the Army's Human Resources Command) should maintain a continuously updated listing of open positions with minimum qualifications. Any qualified individual should be able to express interest in any job.

Each service currently has authority to do this.

6. Expand Information Transparency for Job-Matching

Centralized personnel processes in place constrain information to an extreme degree so that gaining commanders know very little about incoming personnel, and even promotion boards are permitted to see only a fraction of the information available. The current standard is for gaining commanders to be given access to job histories (officer record briefs in the Army), but not performance evaluations or other background. Each service should allow greater transparency and record preservation so that gaining commanders at all levels (division/brigade/battalion) see all possible information on individuals who are inbound or applying to their units. Commanders should be allowed to request additional information to include LinkedIn profiles, letters of recommendation, and communications with references. Likewise, command selection and promotion boards should have broader authority to see this information as well.

Each service currently has authority to make this reform, although legislative language should be simplified.

7. Grant Cyber/Acquisition Workforce Exemptions

The cyber domain has emerged as one of the top threat and battle spaces that conventional military forces were neither aware of nor prepared for a decade ago. Cyber skills are in sudden demand and, like acquisition skills, are ill-served by conventional military personnel rules. United States Cyber Command, currently headquartered at Fort Meade, Maryland, should be granted exemption from the DOPMA standardized "competitive category" career structure as a unique and critical workforce. The same exemption should be granted to active-duty personnel in the acquisition workforce. Exemptions would free members from promotion timetables, tenure requirements, and compensation limits.

Legislative action by the Congress is necessary to authorize this.

8. Increase Tour Lengths

In general, tour lengths should be increased in order to reduce rotation costs and burdens. Congress has already encouraged this action in recent hearings and reports. The services should go further to cement this principle as follows: allow individuals to extend current tour length for one to two years, even on a recurring basis, so long as each extension is approved by the chain of command. This ability would not be allowed for key leadership and development roles and would be limited to ranks above E-4 and O-4. An expanded structure would code each job in the military with a minimum and maximum tenure, to allow more careful career planning by individuals.

Each service currently has authority to do this.

9. Increase Rank Tenure and Career Lengths

Services should be given more flexibility over rank tenure. For example, a service should be able to allow any service member the option to stay at any rank for the remainder of his or her career. This reform would go beyond ending rigid promotion timelines and would, in fact, allow an open ended timeline and longer careers of forty years or longer instead of the current thirty-year cap. The only standards for continuation of service should be competence, performance, and the support of the command chain. To avoid the pre-1941 problem of excessive seniority, all service members would have to continually reapply and be rehired into any billet on a biannual basis.

Legislative action by the Congress is necessary to authorize this.

10. Allow Veterans and Reservists to Apply for Active-Duty Roles

Allow veterans and Reservists to apply for open billets at any rank below general/admiral (O-7). The current lack of permeability eliminates from military jobs millions of fully qualified citizens who have already served honorably on active duty. If any veteran or reservist is physically and occupationally qualified, he or she should be part of the talent pool that the services can access. This would permit lateral reentry limited to honorably discharged veterans, not lateral entry of civilians with no military experience. While reentry of a few individuals occurs under current laws, they are rare exceptions to the rule.

This recommendation raises a larger issue about the structure of the reserves, including active reserve positions and the Inactive Ready Reserve. A modernization of the reserves with an eye toward greater permeability and flexibility is due.

Legislative action by the Congress is necessary to authorize this. This reform in particular requires a careful consideration of pension structure and would be easily enabled by creating a distinct non-pension compensation option.

11. Allow Flexible Sabbaticals

Another kind of permeability can be achieved by allowing active-duty troops to take unpaid sabbaticals. A range of sabbatical options should be available to include (1) nascent programs that contract the individual to return to active status after a set period but also (2) open programs that offer individuals a right of reentry to active status within a set period of time that also amends their year group. Current sabbatical programs tend to be inflexible, and should instead offer maximum control to individuals to have a choice over occupational and geographic preferences, rather than forcing them to precommit to return with uncertainty about those factors.

Each service currently has authority to offer sabbaticals, but full flexibility requires broader reforms to compensation, promotion, and assignment processes. Legislative action by the Congress is necessary to fully implement this.

12. End Selective Service (Registration for Draft)

Eighty-six percent of active-duty troops are opposed to manning the force with conscription. Draft registration became irrelevant in 1973 when the All-Volunteer Force was enacted, but it was maintained in case the AVF failed. President Gerald Ford terminated the program in 1975, but President Carter reestablished it in response to Soviet aggression. The Cold War is over, yet the AVF proved doubters wrong by successfully manning a high-quality force during the past decade of war. It is long past time to recognize the draft is an outdated concept, particularly in light of comprehensive reliance on high-skill human capital in the modern professional military. First enacted in 1917, selective service should be terminated on its hundredth anniversary, saving taxpayers $24.4 million a year and registrants millions of hours of wasted time and other resources. The prospect of a future national emergency that requires conscription should not be ruled out, however, so an emergency infrastructure should be maintained.

The Department of Defense (DOD) should retain a draft reinstatement plan for national emergencies, which would provide for a draft to be implemented if ever necessary.

A proclamation by the president, or legislative action by the Congress, is necessary to authorize this.

13. Use Flexible Ranking in Performance Evaluations

Highly inflated performance evaluations are destructive and common in all fields, but nowhere are they more unbalanced than in the US Army and Air Force. The services should create new performance evaluations that extend the Navy and Marine Corps approach of norming by the primary rater's average. Other flexible approaches can also optimize performance differentiation while avoiding the deleterious effects of forced rankings.

Each service currently has authority to do this.

14. Implement Simple, Multisource (Peer) Evaluations

Military units should utilize peer evaluations of performance for development as well as input to formal performance ratings, awards, and compensation. Nobody knows more about performance than peers, so these evaluations should be designed simply enough to allow the principle of "everyone rates everyone" in a unit. One way to do this is to ask each unit member to circle the names of the top five peers. To avoid bullying, constructive feedback could be offered, but it would be visible only anonymously to the rated individual.

Each service currently has authority to do this.

15. Transform Base Pay from Tenure to Role and Responsibility

The services should use occupational and skill bonus pay much more aggressively. Furthermore, the Pentagon base pay formula should consider ending tenure as a pay criterion altogether and should instead use a role and responsibility supplement for each service member. The role bonus would be comprised of increments for skills and occupation (rather than skills alone). This authority would allow services to compensate the individuals who take on tougher jobs (including command) that involve higher career risk, longer hours, and greater stress. On principle, there is no reason to pay a senior O-3 in an easy job more than a junior O-3 in a demanding job, and the same principle applies to E-5s, O-5s, E-6s, and so on.

Legislative action by the Congress is necessary to fully authorize this, though the services can currently implement some degree of merit pay.

16. Expand Pay Flexibility

The armed forces currently use bonus pay as compensation for certain hazardous, remote, and otherwise challenging jobs, though on a very limited basis. This should be expanded to enable decentralized job-matching in an even manner to prohibit local favoritism. Jobs that remain unfilled after a given time should be paid more using a program that already exists: assignment incentive pay (AIP). Wage flexibility is a core principle in labor markets, which is why challenging jobs in less desirable locations such as deep-sea fishing in Alaska are paid relatively more than others of equal skill. The military practice of ordering individuals to take highly unpopular jobs in unpopular locations is inefficient for the mission and harmful to morale.

In order to fill jobs that are open for long durations, services should establish automatic bonus pay programs that would increase each month. For example, an unfilled requirement could be designated with five priority levels. A priority 3 job, unfilled after sixty days, would automatically include a one-time cash incentive pay (CIP) of $3,000. The CIP could be increased proportionally at 120 and 180 days and could also include additional days of leave. After 180 days, qualifications should be relaxed to allow individuals one rank lower to apply. The use of requirement priority levels is a valuable information signal to troops and managers about which roles are in fact critical to the mission and where priority shortages truly exist.

Each service currently has authority to do this.

17. Pension I: Benefit Payments Should Be Later and Larger

The standard military twenty-year cliff vesting creates a perverse incentive for active-duty troops to immediately leave at the twenty-year point in order to begin drawing their pensions (50 percent of base pay, reduced to 40 percent under Military Compensation and Retirement Modernization Commission reforms). A smarter approach would offer the pension payments starting at age fifty-five or ten years after the retirement cliff vesting date, whichever is sooner. To balance the "lost" funds, monthly payments should be increased proportionally. This adjustment would not affect disability or other payments and should not be applied to active-duty troops with more than ten years of active-duty service.

Legislative action by the Congress is necessary to authorize this.

18. Pension II: Expand Retirement Pension Options

In 2015, Congress enacted a significant reform to the military retirement system that will apply to new service members. The reform reduces the traditional defined benefit (DB) of 50 percent of base pay to 40 percent and adds a supplemental defined contribution. While an important step, the reform's impact is likely to be negligible because no cohort's behavior will be affected for a decade or more—and even so, the twenty-year cliff is unchanged. To fully sever the coercive nature of the DB pension on active-duty troops, entering service members should be given a menu of choices during the first year at their first operational assignment:

- Standard retirement plan (this is the status quo, with 40 percent DB and a small, matched savings asset)
- Fifty-fifty retirement plan (25 percent DB and a medium-size savings asset)
- Full savings plan (large defined contribution plan with a large match)
  Legislative action by the Congress is necessary to authorize this.

19. Allow Training and Education Program Flexibility

A severe constraint on service flexibility is the Joint Federal Travel Regulation (JFTR), because it requires any program that lasts fewer than six months to be compensated with expensive "temporary duty" travel pay. Consequently, the services design most training programs to be longer than six months, requiring service members to incessantly move their permanent base location. This regulation should be amended so that a greater variety of broadening programs can be offered. For example, active-duty troops could have the option to participate in congressional and business internships, or other brief training programs, without per diem reimbursement. To be clear, this would not mandate the end of temporary duty compensation, nor would it mandate any service branch to change its education and training programs. It would simply allow services more options to design/redesign training and education opportunities for members.

Legislative action by the Congress is necessary to authorize this.

20. Conduct Regular Personnel Policy Assessments

The DOD should conduct a regular, transparent assessment of leadership culture and talent management in the armed forces. The goal is to assess organizational features, not personal or unit comparisons. Chapter 1 presents an initial methodology—the Leader/Talent matrix—that serves as a prototype for such an assessment. Systemic reviews of personnel practices should be conducted every four years, alternating between the Quadrennial Defense Reviews (QDRs).

Service chiefs should institute a similar assessment of leadership and management practices in the form of exit surveys of service members upon discharge. The exit survey should include hard-hitting questions that evaluate strengths and weaknesses quantitatively, rather than open-ended questions.

The DOD and services currently have authority to do this.

The Heart of Talent Management

The dilemma for reform is whether one thing can be fixed in isolation. Can a market replace central planning if agents are free to choose but prices are set centrally? Potential reforms are interlinked with one another, but which one is the most fundamental? Which are important but not absolutely critical? To avoid confusion or superficial remedies that eventually fail, it is important to identify the central problem—the heart of the Gordian knot—which is job-matching.

TVF reforms are fundamentally aimed at getting the right people in the right jobs, which is a job-matching optimization process that decentralized markets do exceptionally well. Matching people to jobs is referred to differently by the services as detailing (Navy), assignments (Air Force), or slating (Army). Under the TVF, individual service members would volunteer for jobs first; the personnel commands would screen which volunteers are qualified and then winnow a select group of candidates recommended for hiring. Final hiring authority would shift to unit commanders. Service chiefs would have new authority to establish distinct HR rules for their unique services, including whether to change promotion tables (or not) and how much direct hiring authority would be granted commanders at different levels (i.e., O-7, O-6, O-5). Additional TVF reforms to performance evaluations, promotions, and compensation will be allowed at the discretion of service chiefs. These reforms should help improve the optimization process, both improving readiness and driving down costs.

Figure 3.1 presents a schematic of TVF's basic job-matching process compared to the status quo. In the status quo, personnel commands work with a pool of active-duty members up for assignment, which they slate directly (D) against open positions. This process is incredibly complicated by numerous pressures, including the short-term needs of the service, individual preferences, and career management. Detailers at the personnel centers must juggle all the conflicting pressures to try to match faces with spaces.

The process envisioned under TVF job-matching retains *direct* (D) assignments by personnel commands, but the bulk of assignments would be made using a three-step process. First, individuals *apply* (A): any eligible, qualified individual on active duty can volunteer for an open position in a given unit using an online talent management system. Second, volunteers will *be screened* (B) by managers at the

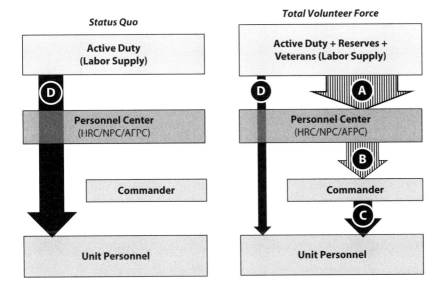

FIGURE 3.1. Job-Matching in the Military: Status Quo versus the Total Volunteer Force

A Individual service members who are eligible and qualified for rotational assignments APPLY to open positions listed in the online talent management system.

B Volunteers will BE SCREENED by managers at the Personnel Command/Center (PC), and three candidates will be recommended to Unit Commanders.

C COMMANDERS will interview candidates and make an offer to their top choice.

D Assignments are made DIRECTLY by managers at PC. This includes promised follow-on assignments and key development positions deemed essential for select individuals.

Source: Author

personnel command, winnowing the volunteers to a list of three or more candidates who are recommended to the unit commander. Third, the unit *commanders* (C) interview the candidates and make a hiring offer to the top individual of their choice. This authority is not new. Rather, it is a restoration of the kind of authority military commanders had during World War II. Nor would hiring authority under the TVF be absolute. Many roles would still be detailed *directly* (D) just as in the status quo, to include priority individuals and individuals who have been guaranteed follow-on assignments.

The goal of all of the TVF reforms is to improve military readiness through the human dimension. Readiness is improved by increasing experience levels and reducing the amount of time spent accessing and training.

A Market Revolution?

The armed forces are different from profit-seeking corporations—the motivations and incentives of its people are different—which makes reform ideas that come from outsiders suspicious in the minds of many. And it is true that command authority and obedience from subordinates are uniquely vital in the profession of arms. The contrast made some military leaders in the 1960s and 1970s hostile to the then-radical idea of eliminating conscription and adopting the All-Volunteer Force. They worried about a mercenary culture. Those fears proved unfounded. The AVF has proved a success in boosting not only the quality, but also the professionalism, readiness, and cohesion of active-duty troops. The market incentives used in recruiting for the AVF were unprecedented before 1973—proactive outreach to young citizens, enlistment bonuses, even commercial advertising—but this market adoption enhanced rather than corrupted military values.

On reflection, Americans recoil at coercion and the tyrannical leader who issues unlawful orders. Absolute (coercive) authority is appropriate on the tactical battlefield, but it cannot compel soldiers to violate rules of war, nor to behave in certain ways outside of the commander's scope of authority (e.g., listen to certain types of music, vote, pray, or spend a certain way). And managerial coercion is grossly inefficient in personnel operations when individuals cannot be required to stay in uniform. Coercion was necessary for a conscripted *citizen* military of short service periods, but not for a *professional* military where enlisted and officer ranks serve for long careers.

Rather than frame personnel reform as a choice between "military" or "market" approaches, we should recognize that military personnel are already in a labor market. It is a planned, internal labor market

that personnel commands are required by law to manage centrally—
and there is nothing especially military about it. The TVF in simplest
terms is a proposal to weed out inefficient, coercive personnel practices
left over from a bygone era. Ironically, today's personnel centralization
was impossible before sophisticated accounting was made possible by
modern technology. Although it became feasible in the 1970s, central management of the military labor force has never been—and can
never be—as efficient as an open market with autonomy for agents of
supply and demand.

The point is that the Human Resources Command and its sister
organizations are currently running the internal labor markets of the
various service branches. No reform will "introduce" markets. Rather,
reform will modify the way the military labor markets work, tweaking
them to make job-matching work better. Reform will allow services to
experiment and modify the various processes in whatever ways best fit
their standards and culture. Marines may want every officer to be able
to serve up to forty years: so be it. The Air Force may want every officer
to attend six weeks of cyber training and be offered unlimited sabbaticals: so be it. The Army may want to reintroduce brevet promotions and
discontinue the use of year groups: so be it. Whatever constraints or
enhancements are tried can only improve systemic learning. But at the
end of the day, the building blocks of talent management are the same:
job-matching (choice), information, and compensation.

Think of available personnel in each branch as its labor supply. Job
openings (usually referred to as manpower requirements) represent
labor demand. Efficiency comes from optimizing the matching of
supply and demand, so that the best-qualified candidates are filling
the open positions. Under conscription, the military treats labor
supply as unlimited, or what economists call perfectly elastic (see
figure 3.2), which gives the military the power to "make" wages extremely low rather than "take" the equilibrium wage set in the private sector. Ironically, under the AVF, the military now treats labor

Under **conscription (before 1973)**, the military treated Labor Supply as perfectly elastic, and acted as a price maker. Military pay was below equilibrium.

Under **AVF (1973–present)**, the military treats Labor Supply as highly inelastic, trying to exactly set Supply to Demand, and must set pay above equilibrium.

Under **TVF**, the military would have access to a naturally shaped Labor Supply that includes troops on sabbatical, reservists, and veterans. Compensation would be closer to equilibrium.

FIGURE 3.2. Three Approaches to Military Labor Supply
Source: Author

supply as something that should be centrally programmed to the exact number, which essentially treats labor supply as perfectly *inelastic,* a driving force in military wage inflation. The goal should be to normalize the labor supply curve so that wages are at a flexible equilibrium. The strategy is to expand the labor pool, which can be done in many ways: (1) allow intra-service flexibility across occupational specialties and age cohorts (e.g., allowing an infantry major to apply for an Army captain's job opening), (2) allow interservice labor flows, (3) allow flexible sabbaticals, (4) allow reservists to apply for active-duty positions, and (5) allow lateral reentry of veterans into active duty.

Why centralization took over the Pentagon is not really the question here, but it must be recognized that concerns about nepotism and fairness were rationales for stripping local commanders of hiring authority. To be sure, the worst of those problems were curtailed by the personnel commands, but all of the positives in a free market were eradicated as well. Information—about which officers were the best, were unique, were laden with potential—was thrown out with the nepotism bathwater, replaced by a one-size-fits-all paradigm borrowed from industrial corporations that saw human labor as interchangeable parts.

Decentralized job-matching will be an improvement. It does not require new technology, only renewed authority for commanders to choose, as well as the authority to interview, to recruit, and to seek out information about candidates. We know from the explosion of Internet businesses such as Craigslist and eBay that matching can be *enhanced* by new IT. However, reform must not be dependent on developing a new IT system. It only requires a devolution of legal authority. Commanders and troops should be given clearance to utilize existing technologies in the hiring process, e.g., the freedom to post and search LinkedIn profiles.

As for nepotism in the TVF, commanders will be held accountable for mission performance, and they will also be held accountable for discrimination (as they are now). Moreover, the TVF hiring process involves personnel centers nominating a limited number of candidates, eliminating the possibility of hiring nepotism by a unit commander. Presumably, the services will want to retain the ability for unit commanders to make "by name" requests as it currently exists.

All recommended TVF reforms presented in this chapter the executive summary are aligned in table 3.1 by category. Later chapters

BOX 1.
Three Fundamentals: Choice, Information, and Price

Three elements are present in any market: *choice, information, and price.*

Economists assume that agents have choice, such as your freedom to choose what to eat for lunch. But the authority to make choices is not always controlled by individual agents. For example, parents make many choices on behalf of their children. In a free market, individuals on the supply-and-demand side of each transaction have autonomy over choice—hiring authority is possessed by bosses, and accepting authority is possessed by workers. In centrally controlled markets, both sides of choice are made by paternalistic coordinators. A *managed* market is a middle ground in which a central authority regulates the degree of choices available. The TVF maintains this management for personnel commanders to screen available jobs and candidates but allows final choice to be decentralized.

Information means what is known (or knowable) about both sides of a market. This includes information about candidates, including application letters, recommendations, phone calls to references, and background checks.

The element of price seems self-evident. In labor markets, price is the wage paid for work plus noncash aspects such as location, hours, vacation, and unit camaraderie, not to mention the personality of the boss. A fun job will typically pay less than a grueling, dirty, monotonous job.

For the standard military market, the price of labor (salary) is not flexible in the way we think of salaries being negotiated in the private sector. The wage is the same for a USAF pilot based in Germany, in Texas, or in Guam. Price inflexibility can cause distortions, but the military avoids the textbook problems of market failure by removing choice from pilots about their assignments and placing it in the hands of the Air Force Personnel Center (AFPC).

In sum, all three elements (choice, information, and price) are constrained in the modern military labor market. Any reform of the Pentagon personnel system needs to be designed to consider each of these three elements. Some reforms to information can enhance the controlled market, but shifts in choice authority will not succeed without richer information networks and some price flexibility.

will delve into research and analysis behind reforms in specific areas. This chapter provides an overview of how the TVF will work.

Core Principles of the TVF

The TVF blueprint offers a number of incremental reforms, but the core concept is the formalization of internal labor markets within each service for officers and enlistees that are optimized for job-matching (best talent to the best job); decentralized, so that commanders have greater control over promotions and assignments; and personalized, so that individual service members and their families are given greater career control. Individuals will be free to volunteer for a wider range of positions vertically (for faster promotion or slower specialization) and horizontally (across specialties and even service branches, as long as qualifications are met). Force-shaping will be done organically. Continuum of service (among active-reserve-Guard-inactive status) will be much more flexible. Table 3.2 makes a comparison between the closed military labor market and an open labor market.

Another key change to make a market work is better information— primarily better performance evaluations that will have real substance: richer competencies, quantitative assessments, and commentary. Likewise, officers and enlistees seeking jobs will have freedom to express interests. Careers will be allowed to slowly shift so restrictive tracks no longer discourage specialization and no longer mandate excessive job rotation. Performance evaluations will be reformed to provide (1) unique information about an individual's skills, character, interests, and potential; (2) distinct metrics that differentiate merit among strong and weak performers across multiple categories; and (3) valuable feedback for leadership development. Developing a new peer evaluation as part of the information system can powerfully enhance talent management.

Compensation will be reformed in three ways. First, the retirement pension will no longer exist as a defined benefit that is earned and col-

TABLE 3.2.

Comparing an Open to a Closed (Military) Labor Market

	Open Labor Market	US Military Closed Labor Market
Choice	Individual authority. Buyer has choice. Seller has choice. Agreement made upon mutual choice.	Central command. No authority to make choices by buyer or seller. Limited ability to express preferences.
Information	Asymmetric and incomplete, but both sides have freedom to learn unlimited information through interviews, references, and direct exchange.	Limited knowledge about quality of individual labor supply. Information considered by central authority is narrow.
Price	Negotiated salary & term. Defined contribution savings plans are norm.	Predetermined salary. Long-term commitment. DB 20-year cliff pension.

Source: Author

lected all at once (what is known as cliff vesting) after twenty years of service. Second, pay and bonuses will be reformed to allow greater flexibility—vital to help commanders fill critical-but-undesirable jobs and to reward volunteers for taking more dangerous, arduous jobs. The standard pay table will be revised so that time-in-service is no longer a factor, replaced by role and responsibility pays. This flexibility is necessary to reward (and thus retain) skills that are in high demand in the private sector. Command roles, as an important example, could be paid a significant markup of base pay, given the high burden of time, pressure, and risk that command entails. It must be noted that the military currently does offer wage variation for occupational skills, but it does so crudely by paying bonuses across the board to some skills but not others while making no wage distinctions within those occupations (e.g., paying all pilots the same flight bonus, so that the worst pilot in the Air Force gets no less than the best, while the best logistician gets no bonus). Third, the rarely used assignment incentive pay (AIP) will be deployed to make job-matching work efficiently by way of using an algorithm to award automatic assignment bonuses for requirements that prove difficult to fill.

Human resource management in all four services is actually two different systems. One generates job definitions and specific requirements across thousands of units (the manpower system). The second manages people and their paperwork (the personnel system). Together, the manpower spaces are matched with personnel faces. The general matching problem can be understood as a two-dimensional grid with thousands of requirements on the vertical and an equal number of people on the horizontal, represented in figure 3.3 as a large square.

A perfectly managed market will generate precisely the right number of people to match the number of manpower requirements, with no oversupply or gaps, and will also optimize their talents with rank, training, and, hopefully, motivation. And it will forecast the needed requirements decades ahead of time so that the personnel system is able to recruit, retain, and grow the correct force with precision in terms of numbers and talents. This planning ideal is, of course, impossible and expensive, as Eric Hanushek observed in 1977, just a few years after the AVF was implemented: "The military services fall into implied contracts

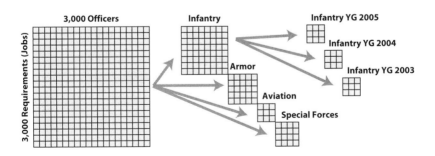

FIGURE 3.3. Breaking Down the Job-Matching Problem
Source: Author

(as with retirement rights), and they must, in general, compensate the individual for assuming absolute control over the jobs and locations in which they can place the individual."[75] Equating the size of the labor pool to the size of manpower needs gives huge leverage to marginal individuals in the labor pool who have a sort of monopoly power to demand higher compensation, which is why military pay has by many measures outstripped civilian pay.

In 2015, there were 10,456 naval officers with the O-4 rank of commander, and somewhat more of the same rank in the Air Force (12,671) and Army (15,692). Roughly one-fifth of these officers are reassigned during each detailing cycle. If the typical officer is detailed to a new position every eighteen months and there are three detailing cycles per year, that means the Army has a matching problem on the order of a $3,000 \times 3,000$ box to solve every cycle. The size of such a problem is too big to be solved centrally, so the military breaks down the big box into smaller boxes using occupational specialties and seniority cohorts (i.e., lineal numbers/year groups). In the abstract example shown in figure 3.3, three thousand Army majors need to be slated against three thousand billets, which is initially broken down into four branches of infantry, armor, aviation, and special forces. The infantry box may have 1,400 officers, which is further broken down into year groups 2003, 2004, and 2005. In reality, there are additional divisions into occupational specialties such as intelligence, logistics, and so forth. But the basic idea is that the computational problem becomes solvable by a human manager when slating dozens rather than thousands. Branch managers (a.k.a. assignment officers) typically focus on distinct blocks, e.g., captains in March, majors in April, and colonels in May.

The TVF solves the problem differently. Instead of discrete assignment periods, the open market relies on decentralized and continuous matching. On any given day, there will be a number of open/opening requirements. A more detailed description will be given in chapter 5, but the basic process has N openings and some multiple $V \times N$ (where

V is greater than 1) of qualified individuals who volunteer. As an example, imagine 500 open requirements for majors and 4,000 Army officers who are currently qualified to apply. Gaining commanders have the benefit of selecting the most qualified of those applicants. Efficiency rises with the multiple V; that is, the object of a system that works best for maximizing talent will have a thicker labor supply.

Figure 3.4 shows how the TVF would deepen the potential supply further by allowing joint officer applicants, sabbatical reentrants, and reservists who can flexibly move into active-duty jobs, as well as lateral reentry of qualified veterans. A qualified veteran would need the requisite military experience, fitness, and clearances in order to be approved by the branch manager. This continuum of service can be made operational fairly quickly by opening active-duty roles to individuals who maintain reserve status.

Now let's imagine a particular case or, rather, set of cases. Suppose there are three open Army requirements for the role of infantry battalion S3 (operations officer). One is *urgent,* meaning the role is either unfilled or will be within thirty days, and it is part of the 1st BCT (brigade combat team) at Fort Drum, New York. The other two will be open in ninety to 180 days. The second job is part of a battalion at Fort Jackson, South Carolina. The third is part of a 25th Infantry Division at Schofield Barracks, Hawaii. A common misunderstanding is that everyone will volunteer for the job in Hawaii, while no one will apply for other roles, but logically only one candidate will be selected

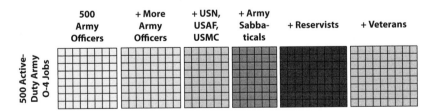

FIGURE 3.4. TVF Job-Matching Approach
Source: Author

FIGURE 3.5. Examples of TVF Job-Matching for Three Battalion S3 Jobs
Source: Author

for the Schofield S3. Let's imagine the volunteers for these open jobs look something like figure 3.5. Drum gets eight volunteers, Jackson gets three, and Schofield gets two dozen, including five reservists willing to go on active duty for the position. Next, the HRC branch manager would screen the volunteers and identify three candidates for recommendation to the gaining battalion commanders. The commanders make their selections, all individuals are notified, and the process continues.

What if nobody applies for the Fort Jackson position? Perhaps the unit is scheduled for an upcoming deployment. Perhaps the base is geographically undesirable. Combat units are surely seen as more, not less, attractive to many officers. And every city and state is able to attract workers for every job imaginable. The insight is that wages rise to compensate for many factors such as geography and danger. This is why the TVF job-matching process should be supplemented with flexible pays. Requirements that are *urgent* (unfilled within thirty days) should offer a supplemental pay of some fixed amount that is not at the commander's discretion of, say, 5 percent of base pay. Those that become *critical* (unfilled after forty-five days) should get a higher supplement of 10 percent, and so on. Different services can refine the supplemental open requirement pays to find what is best for readiness. Service personnel commands might use these pays to signal which roles are truly considered critical by, for example, nixing supplements for unfilled and less important roles.

Currently, military compensation is inflexible given the hard legal structure for pay grades and time-in-service bands. Retirement is also inflexible because of its universal one-size-fits-all nature. However, it is an error to think the military services have no flexibility on compensation, because the law allows much larger bonus pays than any service utilizes. Historically, bonuses are paid to classes of officers and enlistees, but there is no reason they cannot be individualized. For example, bonuses are routinely given for geographic assignments, specialties (flight pay), and hazardous duty. Bonuses are also given for reenlistment (which is standardized) as well as for force-shaping (variable every year, and flexibly applied to different age and specialty cohorts). According to 37 US Code § 332, "Service Secretaries may pay bonuses to officers [who] serve in an active status." A bonus of up to $50,000 can be given on an annual basis. The law is designed to induce individuals to remain on active duty, but there is no reason this cannot be applied flexibly as a bonus for the type of duty, i.e., the specific job and location filled.[76]

What if individuals apply to change jobs every six months, swamping the personnel centers? The TVF would curtail the unlikely but possible chaos of too much job searching by establishing minimum and maximum tenures for each job requirement. Individuals would not be able to actively apply for their next assignments before fulfilling their existing unit obligations.

What if an individual applies to numerous jobs but is not selected by any commander? This is a valid question, but there are two important answers. The first concerns highly valued officers (and noncommissioned officers), where numerous rejections could hurt morale and lead to a retention problem. This seems unlikely, given that individuals with strong records will be highly coveted. The TVF marketplace would, for this very reason, work in two directions, meaning that it would list all available individuals that branch managers and gaining commanders could scan. Commanders could tag individuals, expressing interest, and branch managers could recommend roles. The full system would likely be overflowing with open requirements over a long time horizon. Finally, branch managers will maintain authority to directly fill many jobs specifically to protect the careers of top talent. It is important to remember that the TVF is not anarchy; rather, it is meant to be a *managed* market.

The second answer is that some individuals will not be selected because they have poor performance records and are not attractive hires. This is a feature of the TVF, not a fault. Especially during drawdowns, a job market helps to naturally cull the force without the ugly and often unfair processes involved in the past. However, the downside is that some individuals might face sudden unemployment after ten or fifteen years of service. This would be grossly unfair without a different retirement program. Until a new retirement program fully displaces the twenty-year pension, a proportional early-retirement option should be offered to nonselected individuals with between fifteen and twenty years of service. Here is one solution: first, nonselected individuals will be encouraged to volunteer for lower-ranked jobs or unfilled lower-priority positions, and if no match is made after three months then, second, the individual will be offered a proportional retirement with equivalent medical benefits that will begin at the age of fifty rather than beginning immediately.

Could individuals apply for multiple jobs? Yes. Ideally, they could bid on open requirements by noting the priority of their preferences for personnel commands to consider. For example, an individual might volunteer for all three jobs in our figure 3.5 example. They could even become candidates for two or more jobs and be selected by two commanders. Just as in the "real world," the highly sought individual would choose from among multiple offers, and some unit commanders would move on to alternates. To avoid rigging the system, open requirements would have to be posted for a minimum number of days before candidate selection, with the only exception being sudden, critical openings.

A well-designed online system would allow individuals to set up daily "alert" e-mails for jobs that matched a search profile.

Why Not Pilot Projects?

Many reforms have been piloted or implemented in the past but failed in their first iteration and were terminated, so subsequent reform ideas are met with skepticism. For example, a current pilot project allows active-duty troops to take two-year sabbaticals. This is intended to experiment with flexibility in rigid career timelines. However, the program is itself inflexible by extracting extra service tenure requirements, of fixed duration, and adding career uncertainty upon return. If the sabbatical program fails, it might well be interpreted as a rejection of career flexibility despite being loaded with inflexibility.

For many people who are serving on active duty, these concepts can seem unrealistic. The surprising fact is that many are already in place in one or more of the services right now. The Air Force implemented an internal labor market program known as OVAS (for Officer Volunteer Assignment System) in 1992. It had a mixed record, popular among many young officers but not with many commanders because it allowed choice among airmen (who could apply for jobs) but not among commanders. Even so, OVAS electronic boards listing job choices had limited information and no wage flexibility. As a result, some positions were impossible to fill. Rather than letting the program evolve, the Air Force abruptly curtailed it in 1998 and reasserted a coercive, centralized job-matching process. The Navy experimented with a market in the 2000s known as Project Sail; the Army in the 2010s piloted Green Pages for engineering officers.

Command authority was generally sacrificed to greater central control in the military during the Vietnam era—a trend that was seen as a positive improvement when it was first made possible by computer technology in the 1960s. New advances in IT now allow for greater productivity through decentralization. However, alternatives to the status quo are hamstrung by federal laws, particularly the strictly defined career progression under the military up-or-out dogma. No pilot project can truly break the mold without legislative action.

Consequences

The primary objective of a reformed personnel system is to improve the quality of the active-duty workforce. However, the most immediately observable consequences of great career flexibility will be higher retention, longer careers, and longer time on station. The pace of personnel churn will slow, leading to a demographic shift in the shape of the force. Current military age demographics are published by the Defense Management Data Center. Its most recent report says:

> Nearly one-half (49.6%) of Active Duty enlisted personnel are 25 years of age or younger, with the next largest age group being 26 to 30 years (22.1%), followed by 31 to 35 years (14.0%), 36 to 40 years (8.8%), and those 41 years or older (5.6%). More than one-quarter (25.7%) of Active Duty officers are 41 years of age or older, with the next largest age group being 26 to 30 years (22.5%), followed by 31 to 35 years (20.7%), 36 to 40 years (17.8%), and those 25 years or younger (13.4%). Overall, the average age of the Active Duty force is 28.6 years.

Because the TVF will allow longer-tenure careers of forty-plus years, discourage mass retirement at twenty years of service, dispense with up-or-out in favor of specialization, and allow permeability, the combined demographic effect will flatten the age profile of the force. Figure 3.6 is a conceptual look at how the TVF will reshape the force over the long run. Allowing longer tenure and more flexibility will naturally increase retention rates (and experience levels) over time, which will reduce the number of officer and enlisted accessions that are needed. Most visibly, the incentive (and pressure) for members to retire

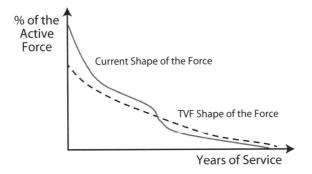

FIGURE 3.6. Shape of the Force
Source: Author

from the ranks at the twentieth year of service will smooth out the current twenty-year preretirement bubble. To be clear: rank structure will not change, but the age structure will. Under the TVF, there will be a wider range of ages at O-4, for example, both younger and older.

The side benefit of TVF reforms and improved readiness will, ironically, be reduced financial costs. Most importantly, longer careers mean fewer new personnel. I calculated that if the number of enlisted service members over the age of forty-one were allowed to increase (e.g., from 31,000 to 55,000 in the Army), then the number of accessions could be reduced by 10 to 15 percent. The services will be able to enhance recruit quality and retain quality, while reducing training and retirement costs with no financial loss to individual service members.

A Strategic Studies Institute report further makes the point that "giving officers greater voice in their assignments increases both employment longevity and productivity," further noting that "failure to do so, however, in large accounts for declining retention rates."[77] A slower pace of churn will initially be interpreted by some as a problem because it will reduce the demand for recruits and reduce the amount of mid-career training per year. That is the consequence of a higher ratio of operational time relative to training time. Military training programs will see smaller and fewer classes, from basic training to War College, yielding cost savings along the way.

As Eric Hanushek observed, "Senior promotions depend upon a wide variety of experience and, therefore, a wide variety of training."[78] The problem of careerist box-checking has worsened since he wrote, mirrored by overtraining. Hanushek noted that it is easy to find "officers who have spent one quarter to one third of their total careers in formal training." Indeed, every mid-rank officer is essentially required to get a master's degree—many of them unwanted and without value to the military mission—and many officers have two or more advanced degrees. This wastes money and, more importantly, time. Increasing the operation/training ratio will enhance readiness but also allow a smaller force to accomplish the same overall mission.

The direct budgetary consequences of reduced churn through the TVF will be much lower permanent change-of-station (PCS) moves. Currently, DOD spends $4.5 billion on six types of PCS moves that every member uses throughout his or her career. In 2016, accession moves are expected to be about 24 percent of all PCS moves, averaging $2,289 per move.[79] Separation moves make up 32 percent of all moves and include retirement, involuntary leaves, and voluntary leaves. TFV reforms would lower accessions (and subsequent separations) by at least 10 percent.[80] Reducing accession and voluntary separations by 10 percent would result in 40,000 fewer moves, saving over $100 million in one fiscal year.

A 2012 Senate Armed Services Committee report acknowledged that some of the biggest personnel savings could be found by extending operational tour lengths, thereby reducing the number of rotational PCS moves. The report states that lengthened tour times would "lead to less stress on the force and hardship on families that are forced to move frequently." It recommends a 10 percent reduction in operational and rotational moves, which would result in $293 million in savings in FY 2013 dollars.[81] Similarly, TVF reforms would make combat

TABLE 3.3.
Potential Savings from PCS Moves

PCS	FY 2016 ($M)	Cost per Move	# of Moves	% Change	TVF Moves	Savings ($M)
Accessions & Separations	$1,079	$2,626	410,978	10.0%	370,000	$108
Operations	$2,854	$11,842	241,036	10.0%	217,000	$285

Source: FY 2017 President's Budget, Military Personnel Programs (M-1); GAO 15-713

tour limits completely voluntary, allowing members and their families to stay in locations they prefer for any amount of time.[82] Table 3.3 shows that overall TVF reforms could save about $400 million per year by reducing PCS moves just to percent.

Operational moves make up 15 percent of all moves and rotational moves are at 18 percent, which together add up to an estimated 241,036 moves. These types of moves, however, make up more than 64 percent of the PCS budget, with the average cost per move ranging from $10,000 to more than $13,000. In FY 2016, $1.12 billion was allotted for operational moves and $1.7 billion for rotational moves. Extending tour lengths would drastically change the number of moves per year. A reduction of 10 percent of operational and rotational moves in 2016 would save $285.4 million in one fiscal year.

An improved performance evaluation process will improve command selection as well as the efficiency of job-matching. A new peer evaluation process will help weed out bad leaders, improve leadership development, and identify diamonds in the rough who are overlooked by the top-down evaluations in the status quo.

Perhaps the great unsung virtue of the TVF that is impossible to quantify is organic force-shaping, allowing natural identification of weak performers who are unable to find jobs internally. A period of downsizing need no longer involve ad hoc programs to reduce the force. The services will be able to adjust manpower requirements in real time, and the TVF job-matching process will winnow the shape of the force automatically. On the other hand, what will happen in the case of a needed surge in force? Here, too, the TVF can respond more nimbly than the status quo. Recall the concern over the quality of enlisted ranks during the Iraq war—the existing personnel system was strained to recruit enough individuals to meet the needs of the services and felt real operational pain by losing more mid-rank noncommissioned officers (NCOs) and officers than projected. The TVF could address force surge better in two ways. First, permeability of service would allow prior service members to seamlessly rejoin the active-duty ranks by design. The only moving piece needed for a market to clear—that is, for higher demand to be met with potential supply—is price. And by design, the TVF is built with automatically increasing assignment bonus pays that scale with need. Second, tour lengths would no longer be of fixed duration and could be voluntarily extended at the discretion of individual service chiefs and unit commanders. Unlike the stop-loss emergency, extended tours would be met through voluntary measures in the TVF. There would be a cost, but it would be shouldered by the military budget, not by involuntary coercion of troops.

Finally, a more flexible personnel system will allow each service chief to optimize the ranks. Unlike today's structure, the Navy and Marine Corps would no longer be required to have the exact same promotion timetables as the Army and Air Force. Innovation is one natural outcome of organizational flexibility. Higher morale is another. Most importantly, the Pentagon will be able to meet new personnel challenges that are over the horizon. Instead of scrambling to grow cyber and drone capabilities as they do today, when the technological ability is hamstrung by manpower limits, the TVF would prove much more nimble in staffing the right people through faster specialization, more flexible training, and better recruitment.

CHAPTER FOUR

Reforming Evaluations

Talent management is impossible without talent evaluation, and that means performance appraisals, also known as fitness or evaluation reports. Unfortunately, performance evaluation does not have a best practice, nor is there much in the way of science to it in private, public, or military organizations. A standard format emerged decades ago, but the only notable benchmarks in the standard annual performance review are that they are (1) widely disliked by managers and workers, (2) widely perceived as ineffective or even counterproductive, and (3) highly inflated. Indeed, the main scholarly insight in the decades-long study of performance evaluations (PEs) is that rating inflation is endemic, multicausal, and a common source of unfairness.[83]

Performance evaluations can be used for two distinct functions: personal development (feedback) or talent management (promotion, compensation, and so on). Often a firm uses a single PE process for both, but it is common for an organization to treat the functions differently using two different systems and sets of forms. This chapter is primarily concerned with the evolution of PE for talent management, which by nature is seen and utilized by managers and the larger organization. Feedback tends to be more private. Yet the lessons in this chapter about why PE tends to fail are applicable to both functions.

Some firms have developed highly effective PE systems—General Electric and Procter & Gamble have been celebrated for decades—but those systems are often ad hoc, using private methodologies. Now it turns out that those systems have caused more problems than managers were willing to admit. According to a Corporate Executive Board survey, more than 90 percent of companies use some kind of rating system to measure performance, and 29 percent use a forced curve (to control inflation).[84] But the survey also found that nearly nine out of ten companies plan to change their PE systems in the near future. In organizations of all kinds, evaluations are not only disliked but distrusted. A 2014 Gallup survey found that 54 percent of the employees in twenty diverse, global companies said they felt that their companies' performance management systems were not effective.[85]

An evaluation system can be ineffective in many ways. Some PE systems force supervisors to identify a fixed number of best and worst performers, despite the common frustration that a team may actually have many top performers and no bad ones. Some PE systems, however, give supervisors unlimited discretion, which allows some managers to give strict ratings that are unaligned with inflated ratings given by other managers in the same company. In general, there is a design tension between too much talent differentiation and too little. My theory is that an organization's peak performance will be highest under a PE system where differentiation is *balanced*, a term that has a specific meaning in this context (see figure 4.1). Even so, it can be an elusive optimization problem in practice.

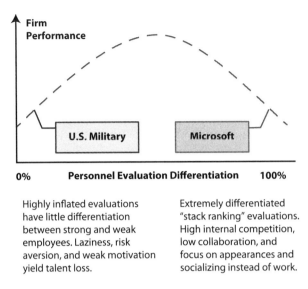

Highly inflated evaluations have little differentiation between strong and weak employees. Laziness, risk aversion, and weak motivation yield talent loss.

Extremely differentiated "stack ranking" evaluations. High internal competition, low collaboration, and focus on appearances and socializing instead of work.

FIGURE 4.1. Optimum Firm Performance at Optimum Differentiation
Source: Author

Inflation without Limits

What causes price inflation? Economists will tell you that the fundamental cause is "too much money chasing too few goods." Rising prices are the inevitable consequence of a government printing more currency: any stimulus to the economy will be short-term, whereas the disruption of expected inflation can lead to long-term chaos. This basic rule of "too much money" causing inflation applies to many other endeavors. For example, online product ratings—including Amazon, Yelp, and Netflix—are vulnerable to inflated scores. This is because users who rate things online face no constraint on the number of five-star ratings. In a sense, raters have "too much money." They can give five stars to a million movies, books, and restaurants, while giving no low ratings whatsoever. With little incentive to rate anything as less than perfect, the subsequent behavior makes ratings inflated and less meaningful.

A similar problem has emerged with inflated grade point averages at college campuses in the United States since the 1970s, spreading to high schools in recent years. Citing data on more than 160 American colleges, Aina Katsikas explains, "The mean GPA for both private and public schools in the 1930s was 2.3, or a C+. That number for both types of institutions increased at the same rate until recently. Today, the average GPA at private universities is 3.3, a B+, while that at public colleges is 3.0."[86] In 2005 Princeton University made a highly publicized effort to curb grade inflation when it asked departments to limit As to just one-third of course grades, but a vote by a faculty committee in 2014 formally abandoned the effort.

Princeton's experience teaches us that inflation is difficult to control once it becomes embedded and that a simple quota of top grades/rankings is ineffective. In that way, Princeton is similar to the Air Force and Army. Even if strict quotas are enforced, rater behavior eventually leads to the full quota maximized and the second tier rating given to all others.

These inflation problems are well documented by decades of management science, a literature reviewed in a comprehensive 2012 article by economists Russell Golman and Sudeep Bhatia:

> Researchers in psychology, accounting and organizational behavior have found that subjective evaluations suffer from severe leniency effects. Performance appraisal ratings display an upward bias . . . with 60 to 70% of those being assessed rated in the top two categories of five-point rating scales. . . . Performance evaluations also are shown to display a centrality bias with supervisors compressing ratings so that they differ little from the norm . . . .[87]

Inflated appraisals happen naturally for a number of understandable reasons. Managers are generally loyal, sympathetic, and fond of members of their team. All but the most cold-blooded supervisors have an inflated perception of the people they lead for fundamental psychological reasons. Given that, ratings are often nudged higher because the manager knows a negative rating puts his team's productivity and morale at risk. The downside risks of authentic PE ratings, personally and professionally, are vastly higher than the potential upside. Because organizations recognize inflationary and centrality tendencies, managers are routinely required to be strict in their evaluations. The conventional method is to establish quotas on top rankings, which almost always conform to the bell-shaped normal distribution. Consider the following two case studies that show how dysfunctional evaluations can be.

Microsoft provides a case study in taking stack ranking to the extreme. When Steve Ballmer took over the CEO role from Bill Gates in January 2000, one of his first decisions was to institute the General Electric–style evaluation system. Jack Welch was the successful and famous chief executive who popularized forced rankings at General Electric. His system required managers to identify their top 20 percent, middle 70 percent, and bottom 10 percent, often referred to as A, B, and C players. This was a useful process for a company trying to get lean, requiring careful supervision, but it may not be effective at other types of firms. A long article in *Vanity Fair* by Kurt Eichenwald[88] described its destructive impact under Ballmer: "Every current and former Microsoft employee I interviewed—*every one*—cited stack ranking as the most destructive process inside of Microsoft, something that drove out untold numbers of employees" (emphasis in original). The Microsoft system required a more detailed ranking than GE; managers had to sort every employee into bins in the distribution: top performers, good, average, below average, and poor. Not only were people stack-ranked inside teams at the software giant, but teams were stack-ranked inside divisions, as were divisions themselves. One Microsoft engineer described its effect: "People responsible for features will openly sabotage other people's efforts. One of the most valuable things I learned was to give the appearance of being courteous while withholding just enough information from colleagues to ensure they didn't get ahead of me on the rankings."

Ten years later, despite dozens of other decisions that may have been strategically brilliant, Microsoft lost its dominant position as a technology pioneer because its PE system had crippled talent inside the company, undermining retention, trust, teamwork, and more. In November 2013, Ballmer announced that Microsoft would end forced distributions; as of today it seems the firm has abandoned formal PE ratings altogether.[89] Yahoo faced a similar dysfunction by using an aggressive version of forced rankings that put it on the far right of the differentiation scale. Even General Electric is phasing out the stacked ranking PE system it used for more than three decades, a system that Jack Welch championed famously but that one scholar now describes as "faith-based."[90]

In 2012, Adobe senior vice president Donna Morris decided to abolish annual performance reviews entirely: no more forms, no more annual meetings. Adobe's internal study found that annual evaluations were viewed as daunting and meaningless by managers and workers and, worse, a detriment to productivity and performance.[91] The standard model proved antithetical to the collaborative and innovative environment the company wants.[92] Adobe's move was quickly emulated by other firms in Silicon Valley and beyond, including Dell and Accenture. "You're Awesome! Firms Scrap Negative Feedback" was a headline in the *Wall Street Journal* in early 2015. The *New Yorker* celebrated "The Push against Performance Reviews" in July 2015. No one knows if the Adobe-led experiment will be successful, but the move shows a widespread lack of confidence in standard evaluations and a rejection of talent differentiation that will move firms toward the far left of the curve in figure 4.1.

Recent research has confirmed that forced distribution PE has significant downsides. A 2009 study by Schleicher and colleagues found that raters in a field study of graduate students found it to be "more difficult and less fair" than traditional, unconstrained numeric ratings.[93] In short, academic scholarship has ample evidence that standard PE tends to be inflated and works poorly, as does the alternative of using forced rankings.

Turning to the US military, we find similar confusion and dysfunction with performance appraisals, though each service has a distinct system. The focus of this chapter is on the use of evaluations for junior and mid-grade officers and non commissioned officers (NCOs), primarily because of the role PE plays in job-matching and promotions. All branches of the armed forces appraise performance for development, not just evaluation, and each service has a wide variety of forms tailored to different ranks (e.g., the Coast Guard uses an A-C variant of the fitness report called FITREP for officer tiers, and all branches use a distinct system for flag rank that tends to eschew rating scales). With that in mind, the armed forces have implemented PE systems that tend to have the worst of both extremes of talent differentiation: forced distribution of some quantitative rankings as well as highly inflated metrics and narratives.

The Leader/Talent matrix assessment (see chapter 1) found that evaluations are one of the weakest categories of talent management for the military (figure 4.2). The category includes four statements, each rated on a four-point scale, from always true (+2.0) to always false (−2.0). The lone positive element is that the military's distinct

A respected program of commendations, medals, and/or awards exists ...
PEs provide valuable information for job assignments and promotions.
PEs make useful distinctions between top, middle, and weak performers.
Assessments of supervisors by peers and subordinates are considered.

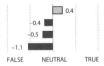

0.4
−0.4
−0.5
−1.1

FALSE   NEUTRAL   TRUE

FIGURE 4.2. Leader/Talent Assessment of Evaluations in the US Military (n = 389)
Source: Author's Leader/Talent survey

culture of medals and ribbons is effective and widely respected. The military PE systems, however, have two severe problems: they are not perceived to provide valuable information for promotions or job assignments, and they are not perceived to make useful differentiation in talent. Finally, peers and subordinates have little or no input in the evaluation process.

To develop a recommendation for an optimal evaluation method, we need to understand the paradox in the evaluation process: inflation in the absence of forced rankings but paralysis in the presence of forced rankings. That paradox haunts most methods that try to force a rankings distribution that is normal in general but not normal in most teams. I posit below that the paradox can be resolved rather easily and the resolution can be understood by examining two of the most effective PE systems in the nation: the US Navy and the US Marine Corps. Those two services use a *rater profile:* each numeric rating for an individual is reported alongside that individual's average rating of his or her rating officer. This profile grounds each individual rating, which in practice maximizes talent differentiation and minimizes inflation without micromanaging raters.

Performance Evaluations in and out of the Military

The annual performance review began with Harvard Business School Professor Elton Mayo in the 1930s and became mandatory for federal workers after the Performance Rating Act of 1950 became law (also setting the standard for the private sector). The idea of appraising performance seems straightforward, but in practice has been unstable and complex because it serves multiple purposes: evaluation as well as development. In the short term, the manager wants to maximize performance, a goal that seems to align with both core purposes, but negative appraisals that can benefit development and help the firm in the long term can hurt short-term morale. The PE serves other purposes, including promotions, compensation, and legal obligations. These purposes have internal tensions and divergent requirements for privacy, frankness, and measurement. Moreover, the entire evaluation process is laden with potential biases, aside from mere favoritism, including leniency, short terms, and conflicting loyalties of the rater (to his subordinates versus the firm, complicated by group dynamics among teams within the firm).

Although this chapter opens with the strong claim that there is no best practice and that performance management is going through a time of uncharted turmoil, a set of norms and standards still exists. Many textbooks for human resource management are in print, with all devoting one or more chapters to performance appraisals. The first lesson is that goal-setting is an essential part of the PE process. Written evaluations also serve as useful documentation, often an essential legal requirement for layoffs. The textbook appraisal has two core pieces: a numeric rating and a narrative assessment of the worker's performance during the past six or twelve months. Typically, the rating section has a five-point scale, with the highest point something like "exceeds expectations"; typically, that scale is applied to three or more categories, such as basic proficiency, interpersonal skills, and leadership potential. In most firms, the rater is not constrained in how many top ratings she or he can award.

The appraisal standard is also described by the Society for Human Resource Management, an unparalleled resource for human resource managers, which in 2012 published an updated Performance Management Standard.[94] The first guideline highlights using a "measurement scale (for example, a five- to seven-point rating scale or a wording scale such as 'does not meet, meets, or exceeds expectations')" and adds that "the rating scale should provide a sufficient number of rating levels to differentiate performance among employees." The second guideline is that a standard PE system "includes consideration of the full array of factors associated with employee performance success. These

BOX 2.
Common Elements in US Military Performance Appraisals

- Evaluation reports are prepared annually for NCOs and officers below flag rank.
- Evaluations are top-down only. The raters are in the supervisory chain. Peers and subordinates are not involved.
- Evaluation forms include narrative blocks for written comments. Each service also includes a distinct set of rating criteria and quantitative scales. Leadership is the only common criterion.
- Three services include an assessment of an officer's promotion potential and a distinct promotion recommendation block.
- Rated individuals meet with their supervisor to be counseled on the appraisal and must sign the report after all reviews are complete. The ratee has a right of appeal.

address the knowledge, skills, abilities and personal characteristics possessed by the employee as they are related to the technical, interpersonal and leadership requirements of the role in question." Whether this standard works better in general, or whether variations are recommended for different types of firms, is unsaid.

The absence of rigorous scholarly guidance may explain why the current US Army Officer Evaluation Report (OER), Army Form 67-10, is the eighteenth revision since World War I. The version it replaced, Form 67-9 (introduced in 1997), was an attempt to fight inflation, just as Form 67-8 (1979) was. Then US Army chief of staff Bernard Rogers cautioned that inflation was a problem but that officers would need to change the ratings culture to provide reliable differentiation for promotion boards. Indeed, it was the introduction of central selection by service-wide promotion boards after Vietnam that made the officer evaluation forms so important to a career, a centralization that subsequently drove inflation. At times, the Army's personnel command was able to stifle inflation by monitoring rater profiles (printed in internal central copies of the completed 67–8), but this norm collapsed during the post–Cold War downsizing when commanders across the Army attempted to protect their subordinates.

A forced ranking of sorts was introduced with Form 67–9 that placed a quota on the top rankings that senior raters were allowed to give. The immediate supervisor (rater) checked one of the boxes with a promotion recommendation (must promote, promote, do not promote, or other) and had no quota. Likewise, the senior rater had no limit on the number of assessments of promotion potential ("best qualified" is almost always checked). But the Army restricted the percentage of relative rankings that a senior rater could give in the section that compares the individual to his or her peers, limiting each senior rater to less than 50 percent in the "above center of mass" (ACOM) group.

In hindsight, it is understandable that a ratings culture developed throughout the US Army of giving ACOM ratings to individuals in the unit who were next up for promotion, not to the top performers. This timing of ratings means that officers commonly oscillate between COM and ACOM ratings on their performance evaluations. Newly assigned officers in a unit almost all get a COM on their OER, then an ACOM the next year when they come up for promotion and another job assignment. The culture has the added benefit of showing improvement in their current roles.

The Air Force has what is likely the most inflated PE system in the world—or maybe in the universe, if we are to use the kind of hyperbole common on AF Form 707, Officer Performance Report (OPR), and the equivalent form for enlisted members (EPR). Like the Army, the Air Force evaluations involve one supervisor/rater and a second, senior rater. Like the Army, many unwritten rules and norms surround what makes an evaluation stand out. Having a boss who is a good writer, knows the lingo, and understands the timing of the system is very important, but having a highly ranked senior rater can trump all else. Thus the timing of getting a high-ranked senior rater is gamed, to the extent possible, and influences promotions.

Because the text blocks in OPRs and EPRs have become effusive, Air Force commanders encourage raters to use stratification and code words to indicate an individual's actual quality. For example, an EPR that recommends the individual be "promoted immediately" might actually hint something sinister, whereas the words "step promote" are understood as a mark of excellence. Over time, these code words evolve, but the awareness and interest of commanders vary widely. Even worse, different code words are encouraged or discouraged by different senior commanders, so that an OPR in one geographic region that was written to signal excellence can easily fall flat a few years later in a different theater. This complexity is compounded for members who have joint assignments.

Numeric assessments on a scale of one to five have historically been a lengthy portion of the Air Force EPR. But because raters faced no quota, giving ratings of "five" in every category (known as "firewall fives") was all but universal except for the rare EPR of an individual being encouraged to separate from active duty by her or his supervisor. A new enlisted rating system was introduced in early 2015 that imposes a quota on the top two rating tiers, similar to the Navy's FITREP, but it is too early to assess its impact.[95]

Air Force members know their evaluation system is dysfunctional but may not appreciate the extent to which it corrodes command integrity. The core military culture values personal integrity above all else, but the Air Force evaluation process requires raters to give fundamentally dishonest appraisals in the form of extremely laudatory narratives for the mediocre and even the least-productive officers and enlisted staffers. The use of euphemisms in PE narratives creates a sense among the force that no official statement should be trusted at face value.

TABLE 4.1.
US Military Performance Evaluations[96]

	Quantitative Distribution Forced?	Quantitative Inflation?	Narrative Inflation?
Air Force "AF 707 OPR"	No. However, raters are encouraged to use stratification in the narrative block—e.g., "No. 1 of 7 captains."	Yes. Unlimited top ratings for FPRs make the numerical aspect of evaluations mostly meaningless.	Yes. Language is gratuitously positive. Culture of coded signals in text is inconsistent over time and among units.
Army "DA-67–10 OER"	Yes. No more than 50% of ratees can be given top rating (ACOM).	Yes. Some categories have unlimited maximum ratings in which everyone must get to advance. Low ratings are extremely rare.	Some.
Navy "NAVPERS 1610/2 FITREP"	Yes. Limited promotion recommendations per rater and organization.	No. Rater profile (average) on every fitness report means that rating is given context.	Some.
Marines "NAVMC 10835 FITREP"	No.	No. Rater profile (average) on every evaluation and strict profile tracking have eliminated inflation.	No. Plain language is used.
Coast Guard "CG-5310 OER"	No.	Yes, but not extrem. The 1–7 scale marks tend to be concentrated in 5–6 range.	Some. Like others, language is used to signal real assessments, and white space reflects poorly.

This table focuses on officer forms and is not exhaustive.
Source: Author

The Department of the Navy offers a bracing alternative in both naval and marine fitness reports, or FITREPs, their colloquial term for evaluations. A comparison of key factors across active duty forces including the Coast Guard is presented in table 4.1. From the theoretical perspective in figure 4.1, the form NAVMC 10835E, Marine Corps Evaluation Form, is just about ideal. It includes thirteen attributes, each marked on a scale of A through G, where A is lowest (considered unacceptable, requiring written justification on the form) and G is highest (F and G also require written justification). Stephane Wolfgeher[97] summarized what makes the system unique: "[Raters] develop a grading history over time (RS profile) that allows for a relative value of an officer's performance . . . a dynamic tool that cannot be reset. The [reviewing officer] also develops a comparative assessment profile. A master brief sheet fitness report listing will evaluate a ratee's received marks in relation to the RS and RO profiles. This evaluation includes the ratee's marks vs. the profile at processing (a static number/value) and the ratee's marks vs. the RS/RO profiles at master brief sheet processing. This last value is a dynamic value that continues to change as the RS/RO accomplishes more and more evaluations. This influences consistent and accurate evaluations as evaluations today affect the relative value of evaluations previously accomplished and those yet to be written." To help guard against inflation, not only are rater (RS) profiles carefully monitored, but the RS is generally rated by the RO using an additional category on fulfillment of evaluation responsibilities that expressly penalizes inflated markings.

Navy FITREPs use an almost identical process for numeric ratings. Another feature of Navy PE is that FITREPs are conducted en masse throughout the force, whereas other services tend to conduct them on a unit or individual basis, depending on commander and member rotations. For example, all Navy O-2s receive FITREPs in February, O-3s in January, and so on. Seven attributes are marked on a five-point scale, with five being the highest. A processed FITREP includes two boxes along the bottom of the form: one documenting the ratee's Member Trait Average and the other the Summary Group Average of other comparable officers assessed by the senior rater. Narrative text also uses stratification comments, but the rating profile keeps inflation out of the system. The downside of the Navy approach is that FITREPs also include a check block for promotion, with a strict quota on how many Early Promotes and Must Promotes are allowed per rater: 20 percent for the former and 50 percent for both.

Talent Distribution in Theory versus Reality

In theory, there is a normal distribution of talent in a large population known as a bell-shaped curve. In a roughly normal distribution, we see talent measured along the horizontal axis, and the frequency of people with a certain talent level is measured along the vertical axis (see figure 4.3). In a normal distribution, 68 percent of the population exists within one standard deviation of the average, meaning that less than one-sixth of the population is very talented (above one standard deviation) and that one in fifty is extremely talented (above two standard deviations). The normal distribution occurs naturally and is found in many characteristics of large human populations, including height, weight, blood pressure, and intelligence.

A principal goal of performance evaluations is for the organization to identify its most talented workers (expressly stated in federal government regulations) to reward them, incentivize performance among all workers, and identify potential leaders. A related goal is to identify weak performers to remedy their problems by retraining, relocating, or removing them. Of the two goals, finding "superstar" leaders is the most vital task, according to academic research indicating that the leadership quality of chief executives can be important for organizational success. Every soldier understands that a superior military force will lose if it executes a faulty strategy; the

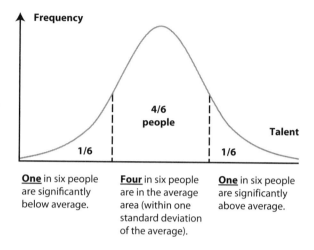

One in six people are significantly below average.

Four in six people are in the average area (within one standard deviation of the average).

One in six people are significantly above average.

FIGURE 4.3. The Normal "Bell-Shaped" Distribution of Talent
Source: Author

corollary is that strategic genius is a key to victory and often survival. Identifying genius is arguably a higher goal for military organizations that face the highest stakes.

The US Army's Officer Evaluation Report was designed to identify top, middle, and bottom performers using a center of mass (COM) designation. In practice, almost no officers are designated in the lowest group (below COM or BCOM). The Army requires raters to limit the percentage in the top designation (above COM or ACOM), which has led to the predictable outcome that barely under 50 percent of OERs are rated ACOM (see figure 4.4). Not only are superstars impossible to distinguish among the Army's ACOMs, but the dividing line between COM/ACOM ratings is placed at the most difficult and heavily populated median of the population, where measurement error is likely to be greatest.

To be effective, a forced ranking system should afford raters the correctly sized buckets to match the shape of their team's performance distribution. For example, if 10 percent of a commander's team

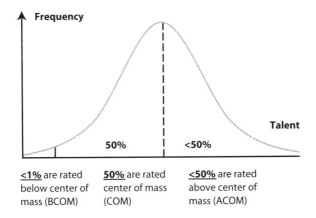

FIGURE 4.4. The US Army's OER Division Across a Normal Talent Distribution
Source: Author's rendition of Army regulations

is clearly in the bottom but she is required to identify 20 percent in the bottom, everyone loses: the team, the boss, and the larger organization. If those numbers are reversed (20 percent bad, 10 percent bucket), the outcome is also destructive to the higher goal of managing people effectively. The moral of this story is that *an effective PE system should have the flexibility to let the rater fit her or his evaluations to the shape of her or his talent.*

In reality, teams with fewer than fifty individuals are unlikely to have talent that fits a normal distribution. A provocative academic paper published in 2012 by Ernest O'Boyle and Herman Aguinis[98] called "The Best and the Rest" asserted that talent is distributed more like a triangle-shaped Paretian distribution than a bell-shaped normal curve. The authors identified a consistent superstar phenomenon in five different fields using empirical data, from professional basketball to academic research. The implication of the research is that a majority of workers are very weak performers, possibly adding zero net value. A careful examination of the data, however, reveals that the O'Boyle-Aguinis case is flawed. Consider, by way of explanation, the performance of pro basketball players.

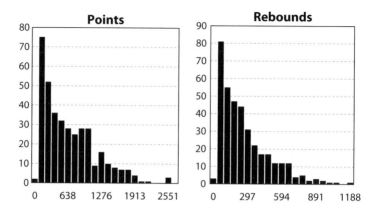

FIGURE 4.5. Data on Professional Basketball Players' Performance, 1981–82 ($n = 393$). Frequency on vertical axis; season total points (left) and rebounds (right) on horizontal axis. *Source:* Author's calculations based on data from basketball-reference.com.

As a random case study, I pulled all player statistics for the 1981–82 National Basketball Association (NBA) season, a total of 373 player records, including Julius Erving and Moses Malone.[99] Larry Bird and Magic Johnson had recently joined the pros, but neither was among the top ten scorers in the league. A histogram of total points scored by each player reveals a distinct Paretian distribution (figure 4.5).

Three players were in a class by themselves that season. George Gervin set the maximum at 2,551, followed closely by Malone and Adrian Dantley. Only two other players were within the top quartile bins. The problem, however, is that total points are not actually how coaches assess basketball players. We know that other factors matter just as much, even if we focus on offense: assists and rebounds. That observation confirms that an efficient evaluation considers *multiple dimensions*.

It turns out that every quantifiable aspect of talent also follows a Paretian distribution. Consider total rebounds. Most NBA players get a third or fewer rebounds than superstars get. A quarter of all players get 10 percent of the top rebound level. Again, this is a Paretian distribution, or what seems to be a slam-dunk for the O'Boyle-Aguinis thesis. That simple sum, however, is not how coaches evaluate talent.

Coaches focus on *efficiency*: how much performance is achieved *per minute*. No player would be considered a bottom talent if he only gets to play a few games in an eighty-two-game season. Once we convert the points data into *points per minute*, we see a very different distribution: a normal distribution (figure 4.6).

That should give human resources executives some relief in that their models are not completely miscalibrated. Yet not all talents are normally distributed—even over large populations—which means some of the core assumptions of forced ranking are probably flawed. Consider rebounds per minute among players in the NBA, which, somewhat surprisingly, have no normal or systematic distribution of any kind.

Keep in mind, these NBA data reveal a performance distribution of 373 players. Most organizational units have far fewer individuals, often a dozen or so. An Air Force wing has hundreds, but a given

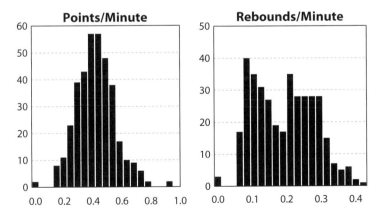

FIGURE 4.6. Data on Professional Basketball Players' Performance, 1981–82 (*N* = 393).
Frequency on vertical axis; points (left) and rebounds (right) per minute on horizontal axis.
Source: Author's calculations based on data from basketball-reference.com

squadron has thirty or so. In smaller groups, it is impractical to assume that one can (1) identify the right talent to measure, (2) measure that talent correctly, and (3) rate people fairly according to a forced distribution of two, three, four, or more bins. The point is that talent is approximately normal but only under extremely favorable assumptions, on average, in large populations. In the real world, team talent is abnormally distributed and will change shape from one time period to the next, especially in military units that fully rotate personnel every two to three years.

Consider three teams, each with ten workers of varying talent on a five-point scale (level five is a superstar, level one is a nonperformer) (see figure 4.7). Compared to three of the forced rankings described earlier—Army, Navy, GE—it is clear that none of the teams could be sorted and ranked fairly by their leaders using any of the three methods.

Using the Army center-of-mass method, team Alpha would unfairly give one average person a top rating and lump the others with the weakest member of the team. Teams Beta and Gamma would face similar inequities but also lose the distinction between superstars and merely great workers. The Navy system of three tiers fares even worse, artificially promoting members in all three scenarios. Finally, the GE system does the best job of identifying the weakest members, but note that none of the example teams have nonperformers. Worse, the GE method fails to recognize the right number of superstars in teams Alpha and Beta. In every real-world forced ranking, an incorrect number of workers is recognized at the correct relative ranking level. Nine out of nine scenarios are failures.

One last thought regarding these three abstract teams: the average talent level is 3.3 for team Alpha, the same as Beta and Gamma. Imagine a team with a majority of very good workers and just a few average workers. Although this is a common perception by team managers, any forced ranking system requires a number of very good workers to be categorized as average or worse.

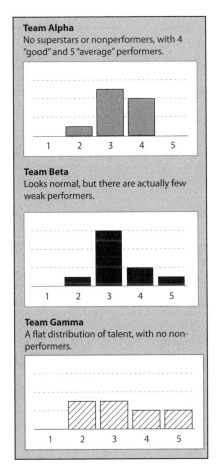
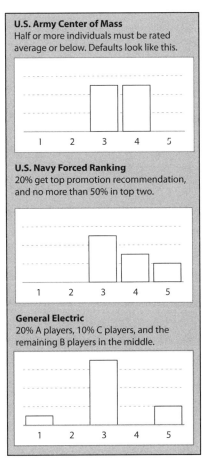

FIGURE 4.7. Talent Distribution in Three Abstract Teams and Forced Distributions in Three Real-World PE Systems

Source: Author's rendition

The vital insight is that no degree of forced ranking will achieve optimum differentiation of talent: not a strict, multitiered approach (like Microsoft in the 2000s), not a moderate number of fixed tiers (like GE and the Navy), and not a binary approach (like the Army). To find the balance between too much and too little differentiation, managers need a method of *flexible* ranking, which is distinct from forced ranking in the same way a cup of water is distinct from a cup of ice cubes. The US Marine Corps allows its leaders to evaluate talent using flexible rankings, even allowing a rater to give none, one, or two of his Marines the very top rating as long as those are balanced against other ratings given to his unit.

The most flexible ranking system that constrains overall inflation allows a supervisor a set number of points to be distributed among members of his or her team. This approach utilizes the economic principle of scarcity. Points are proportional to the number of rated people on the team. For example, a ratio of 3.3 points per person on a ten-person team yields thirty-three points that a manager could allocate among them. Under this method, the manager could theoretically rate the top individual with five points, six points, or more. Indeed, each of the abstract teams presented above—Alpha, Beta, and Gamma—would be perfectly rated using this system.

This section establishes the principle that *an effective PE system should have the flexibility to let the rater fit his or her evaluations to the shape of the talent.* A number of different methods could implement the principle using flexible rankings, of which scarce points is one and mean enforcement is another.

The Need for Peer Evaluations

Over time, PE systems have become labor intensive, requiring many hours of managerial training and many more hours of implementation in creating, negotiating, and explaining the ratings. An internal review by the consulting firm Deloitte found that the ratings of its 65,000 workers took a combined two million hours a year.[100] Despite the large time requirement, managerial appraisals often fail to measure performance accurately for an additional reason not mentioned previously: managers have limited perspectives. A seminal study of PE quality was published in the *Journal of Applied Psychology* in 2000 that reported that 62 percent of the variance in ratings was due to rater perceptions (measured by comparing two or more raters of the same worker), whereas actual performance explained 21 percent.

Those with military experience understand that a commander has a narrow perspective of individual effort, talent, and potential. Commanders are equipped with a staff of subordinate officers and NCOs to assist a unit for precisely this reason. To be sure, each member of a team has only a narrow perspective on the performance of other individuals, but mathematically the best appraisal is a sum of as many perspectives as possible.

The traditional military unit is managed using a strictly enforced hierarchy, known as the chain of command. This structure, and the top-down nature of traditional military PE, has a predictable effect on behavioral incentives. Some subordinates can and do advance their careers by focusing excessively on making a good impression on their rating commander while being otherwise toxic toward subordinates and peers (as long as that toxicity is not perceived by the commander). As Lieutenant Colonel Timothy R. Reese wrote in a 2002 Army War College paper, "Bosses are often fooled by the sycophant or bully— peers and subordinates are not so easily taken in."[101]

The US Army Ranger School uses peer evaluations as an integral part of the training program. Individuals are routinely removed if rated poorly by a wide group of peers in the same class. The rationale is that peers have insights on the quality of an individual's teamwork, attitude, effort, and potential that trainers lack. Students offer (and receive) peer evaluations three times during the course, which involves rank-ordering fellow students and answering two basic questions: "Would you go to war with this person?" and "Would you share a foxhole with this person?" If a student is "peered out" with a majority of negative ratings, that individual is recycled to a different platoon for the next phase and given a second chance. The consequences of ranger peering are a central shaping event in the lives of elite Army soldiers and are widely cited as a key tool in promoting excellence. Why then does peering play no role in standard Army performance management?

To rectify the common problems with top-down appraisals, academics and human resource professionals have been developing multisource assessments that include feedback from colleagues above, below, lateral, and even outside the organization/team of the rated individual. What has emerged is a method known as a 360-degree review that asks a small sample of peers and subordinates text-based questions that focus on the interpersonal skills of the rated individual. In 2015, Rand issued an eighty-eight-page report[102] that affirmed 360s are most commonly used as a feedback and development tool, not PEs, for promotions, job-matching, compensation, or recognition. Employees have generally favorable impressions of 360s, and most senior managers appreciate their benefits.

Rand reports that each of the armed forces uses 360s in some capacity, especially for general officers such as the Air Force's Leadership Mirror and the Navy's one-week program for new flag officers. Those programs are distinct from, although overlapping in some ways, command climate surveys that are also used by all the services. The time requirements of 360s are daunting, however, and they have other costs and limitations. Eliciting honest responses is a challenge; studies show that inflated appraisals increase when 360s are used for anything but private development.

A new Army-wide 360 program (the Multi-Source Assessment and Feedback program) is required for all officers once every three years. It tries to control for a biased selection of raters, a common weakness in the standard approach. In most military programs, the participants select their raters, which is rife with perverse incentives. In short, the new Army program could be better designed.

At the conclusion of its report, Rand recommends limited adoption of 360s by the US military for leadership development purposes only. Rand advises against using 360s for evaluation purposes, but only the narrow method it considered: the narrative-heavy, ratee-selected, small sample model known as the "360." Rand, like many others, made no effort imagining a better method. So let's be careful not to confuse the concept of multisource appraisals with the orthodox 360.

In principle, an effective peer evaluation will adhere to four principles. One, everyone rates everyone. Two, the rating method is simple and fast. Three, ratings are done anonymously. Fourth, individuals are assessed on multiple dimensions. Army Ranger peer ratings meet all four principles.

The first principle does not require each individual in a unit to literally conduct an appraisal of every other person; rather it can be satisfied by giving each individual the *opportunity* to appraise every other person. For example, the rating system could ask each team member to simply select/write/circle the names of the five other team members who have contributed the most to mission success during the past year. As for multiple dimensions, qualities that will be of interest to most organizations include productivity, integrity, leadership potential, work effort, and agreeableness. Each different organization will want to ask about two or three factors most relevant to its culture. In the military, with a culture of constant promotions, asking peers about which top members in a unit are ready for promotion would make sense.

Recommendation: Team Rating

The challenge military services have in developing a better performance evaluation process can be overcome by developing an innovative multisource review that is based on sound economic principles.

I had hoped to find a solution in the academic literature or private sector and was surprised to find nothing. I believe the various branches can easily develop and implement a regular (quarterly) multisource evaluation that is easy to use, fast, and positive and that gives deeply meaningful assessments for use in personal development and talent management. I call the method Team Rating.

Team Rating is based on the principle that everyone rates everyone. Each team member is asked to anonymously select the top individuals in the unit on three different dimensions. The dimensions that would be most useful for military units are: (1) value to the unit, (2) work ethic, and (3) leadership and promotion potential. Larger units would be encouraged to identify top performers in different cohorts; a natural cohort division is officer/enlisted/civilian.

For example, imagine a Navy unit conducting a top 20 percent Team Rating. Each individual would be asked to identify the top two of ten officers, top ten of forty-nine enlistees, and top three of thirteen civilians across the three dimensions of value, work ethic, and leadership. A sailor could be given a sheet of paper that lists everyone's name, followed by three columns of checkboxes for each dimension, or could conduct the rating online. In either case, the sailor selects five officers in the value category, but might choose a totally different five officers in the work ethic category, or might select some officers as being in the top five in all three categories.

A Unit Report will be generated for the unit leadership. The anonymous peer ratings would be tallied and prepared to include distribution for each cohort, by each cohort, across each dimension. For example, the rating of ten unit officers by forty-nine enlistees would look something like table 4.2. Notice that these anonymous ratings reveal only two officers to be far above average in all dimensions—Olds and McArdle—even though there are many officers who are outstanding in one or two areas. The second-highest rated officer in

TABLE 4.2.
Team Rating Example

Name	Value to Unit	Work Ethic	Leadership
Shawn Olds	28	23	26
Kate Carter	17	14	0
Martha McArdle	14	11	22
Troy Thomas	10	17	0
Dave Urban	7	5	14
Daniel Beldy	7	1	13
Emily Yakima	5	5	7
Jonah Casebeer	4	7	10
Chesca Gilroy	4	0	6
John Watson	2	15	0
TOTAL	98	98	98
Average	9.8	9.8	9.8

Source: Author

terms of value is Carter, even though this individual received zero top marks in the dimension of leadership. In contrast, Beldy was given high marks in leadership potential, despite a low amount of feedback in work ethic. In this example, a truly mediocre officer would get an average of ten points in all three dimensions. The reality is that nobody is average, and this method reflects that.

A personal report will be generated for each individual participant. This report will include the tally of top marks for the individual in each dimension, compared to the unit median. When fully developed, it will include anonymous and private comments from peers as well. If there is more than one cohort, multiple tallies will be included in the report—that is, a tally of officers and a separate tally of enlistees.

Conclusion

In conclusion, the US armed forces currently have the freedom to design and use performance evaluations unique to their service needs. This flexibility has yielded two world-class systems in the Navy and Marine Corps that are models of optimal talent differentiation. It is strongly recommended that each military service continue to have the freedom to design its own system. With that in mind, there is tremendous potential to improve how the military branches use their performance evaluations and other information in all of their personnel functions. In addition, all services could—and I believe should—develop multisource evaluations, an area that is not well developed in the private sector.

This chapter has described the two common failures of most PE systems: inflation and forced rankings that do not fit varied talent distributions. It also describes an innovative approach to multisource (peer) evaluations that goes beyond the orthodox 360 evaluations used with mixed results in the private sector.

An ideal evaluation system involves flexible rankings across multiple dimensions. The current Marine Corps FITREP is the ideal evaluation. The evaluation used for roughly a decade by Microsoft Corporation violated both principles and set a standard for failure, showing how a poor PE design can cripple an otherwise vibrant organization.

None of the services uses peer or subordinate appraisals as a formal or even informational input into evaluation reports. This is a missed opportunity. The services are strongly encouraged to experiment with and implement simple, broad, multisource evaluations. A constrained system of top-rating 20 percent of individuals in the unit would be easy to implement on a pilot basis. Peer evaluations of performance should be used for development as well as input to formal performance ratings, awards, and compensation. For development purposes, constructive, anonymous feedback should be solicited that is visible only to the rated individual.

CHAPTER FIVE

Reforming Promotions and Assignments

You are a US Navy fixed-wing attack squadron commander. Your squadron includes roughly 250 personnel, including one other commander (O-5) who serves as the executive officer (XO), four to six lieutenant commanders (O-4s) in the department head (DH) roles, and a dozen lieutenants (O-3s). The squadron includes eighteen aviators, with scores of sailors spread across the four main departments: operations, maintenance, safety, and administration. Yet, like other commanders of military units from Army battalions, Air Force squadrons, and Navy destroyers, you have no hiring authority. You cannot hire, directly or indirectly, any of the individuals under your command.

As the commanding officer, you are most concerned about the talents, readiness, and judgment of your department heads. You can communicate desires for certain individuals to Naval Personnel Command (NPC) detailers, but by-name requests are rarely considered originating from anyone other than a flag-rank officer. A commander also has limited power of negation to discourage detailing of specific individuals. Once someone is slotted, negation is almost impossible without creating a gap in the billet, meaning that the billet will remain unfilled for a year or more.

Imagine instead that you are a battalion commander in the Army. Here, too, you have extremely limited hiring authority, at least outside of the larger brigade. The typical battalion includes a lieutenant colonel (O-5) in command, two majors (the S-3 operations officer and the XO executive officer), ten captains, three or four dozen lieutenants, and 700 enlistees. Depending on the brigade, battalion commanders may have some say about which officers slated by Human Resources Command (HRC) to the brigade are internally assigned to their battalions. Many brigade commanders do not involve battalions in the sorting. Generally, proactive involvement by battalion commanders is rare.

This raises the question: why does *this* batch of majors get assigned to *this* brigade?

The initial slating of the two key leadership roles in the battalion—the S-3 and XO—are handled through division-level recruiting of majors from the larger Army. The majority of majors are in the process of graduating out of the Army's mid-career leadership school—the ILE/CGSC training program at Fort Leavenworth, Kansas. Division G-1s and chiefs of staff travel during the year to Leavenworth before ILE graduation to recruit. For some career specialties, division representatives meet to sort and slate the majors. This process is an impressive utilization of engaging labor demand for sorting, but it has a number of flaws in practice. First, many units are not co-located geographically with Division HQ and tend to be neglected entirely. Second, many O-4s in the Army who are not at CGSC are neglected, including those at different ILE programs, which are sometimes more selective.

You may be the rare battalion commander who will proactively contact HRC to informally search for potentially strong O-4 candidates Army-wide. You might even reach out to officers directly to solicit their interest. But even that does not guarantee the by-name request will be filled by branch managers at HRC.

Captains are assigned to brigades, and members of each new crop of O-3s are doled out to battalions based on the discretion of the brigade leadership. For example, if the brigade has six open spaces for military intelligence junior officers, it will be sent six intelligence captains that cycle. Cycles occur twice per year (previously three times). Often it is clear who the superstar among the group is, but a battalion commander rarely does a background check on the others to petition for select individuals to be assigned to his unit.

The background information available on officers slated to the brigade is limited. Specifically, that means gaining commanders are able to see the assignment history, known as the Officer Record Brief (ORB), but are not able to see any performance evaluations. The consequence is that high performers can be invisible if their assignment records appear normal, whereas a toxic or erratic performer can appear superior.

Why are most commanders not proactive in searching, diligently reviewing, and requesting key staff? First, many battalion commanders are unaware of their limited authority to do so. Second, proactivity runs against the Army cultural norm, which is to accept all assigned troops as equals. Your commanding general might not like seeing you manipulating the system, and why risk his displeasure? Third, trying to manage the personnel system is a diversion of a commander's attention from what the Army emphasizes, which is operational leadership.

The authority to fire and reshuffle individuals out of an Army unit exists but is rarely utilized. There is zero authority to reject assigned individuals and a very limited ability (rarely used) to petition against one's assignment to a battalion. Firing or even rotating out officers who are underperforming is a career killer for them, so they are usually allowed to stay in a role for at least twelve months even when grossly incompetent.

What's clear when we think about military assignments in detail is that there are three systems, not one, involved. The first is the *manpower* system defining the job requirements throughout the service. The second is the *personnel* system that develops and prepares individual service members to fit the required shape of the force, done by managing careers so that the right number of officers in each rank and specialty block is available at the right time. The personnel system also has the unenviable task of centrally matching individuals to vacant requirements. The third system is *promotions*, also centralized, overlaid on the personnel structure.

This chapter recommends a number of reforms to military promotions and assignments, first reviewing problems highlighted above in the status quo. The reforms are not intended for all ranks, however. Initial assignment out of basic training, to include occupational specialization, is beyond the scope of this review, nor have I even attempted to discern inefficiencies. Rather, the adoption of TVF principles should be focused on the middle ranks, namely E-5s through E-7s and O-3s through O-5s. I would caution that the mid-career period is where reform is most needed and that implementation will be problematic for entry-level individuals getting their first assignment. Likewise, there are carefully considered programs for senior officer selection that are more personalized. A central premise of the TVF is that personalized attention is what is lacking, and is arguably impossible to centrally manage, in the middle ranks of the larger populations in the Army, Air Force, and Navy.

Leader/Talent

The Leader/Talent matrix assessment (see chapter 1) found that job-matching and promotions are two of the weakest categories of talent management in the military (figures 5.1 and 5.2). These two categories include four and five elements, respectively, rated on a four-point scale from "always true" (+2.0) to "always false" (−2.0).

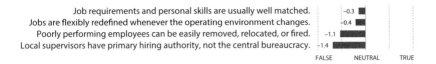

FIGURE 5.1. Job-Matching (Local Control, Efficiency, and Removal)
Source: Author's Leader/Talent survey

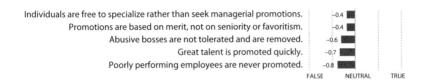

FIGURE 5.2. Promotions (Merit, Differentiation, and Specialization)
Source: Author's Leader/Talent survey

None of the nine elements were rated positively by the panel of active-duty and veteran respondents. By far, the lowest scores were for two aspects of job-matching. The lowest score was for the negligible hiring authority that unit commanders have, which is in stark contrast with nonmilitary organizations (score −1.4). The second lowest score reflects the inability of unit commanders to easily remove, relocate, or fire poorly performing employees (score −1.1). Moreover, the manpower system was assessed as not dynamic and unable to quickly redefine job requirements to reflect a fluid operating environment. These results suggest that the personnel system is not structurally designed to handle new threats and opportunities. Whatever the cause may be, the bottom-line assessment is that the military does not usually match skills with jobs very well.

Regarding promotions, the Leader/Talent assessment revealed negative scores for both the quick promotion of great talent (−0.7) and the nonpromotion of poorly performing employees (−0.8). In fact, feedback indicates that abusive officers and senior NCOs are tolerated in the armed forces. There were less negative, but still negative, scores regarding the consideration of merit in promotions as well as the ability of individuals to specialize (both −0.4). To be clear, these negative scores stand out in contrast with positive scores that respondents gave to other aspects of the military talent management. For example, in the category of development, respondents gave +0.7 to statements about the presence of great leaders in the organization and +0.7 that young leaders are given serious responsibilities. The point is that these responses are not negative across the board. Rather, the military personnel structure is particularly bad at promotions and job-matching.

Given the central importance of these functions to personnel operations, the TVF offers a wide range of recommendations. Many involve changes to the rules (and often laws) governing assignments such as "competitive categories" and "promotion zones," but TVF also recommends the development of a true marketplace that would be a fundamental change from the centrally managed process in place today. It should be noted that the current personnel rules create a straitjacket on HR processes that yields a one-size-fits-all force shape for the four different branches, as if the Marines are optimized with the same rank pyramid and career timetables as the Air Force. Changing the rules will allow service diversity. Adopting a marketplace will enhance matching optimization (i.e., productivity), regardless of the rules.

The Roots of Personnel Inefficiency

For most of its history, the United States military was haunted by seniority. Perhaps the most extreme example came after the Civil War when a large cohort of naval officers held onto senior and even middle ranks—refusing to retire—causing a severe shortage of promotion opportunities for younger officers. Top graduates of the

Naval Academy's class of 1868 remained lieutenants for twenty-one years.[103]

The Army and Navy attacked this problem in different ways, first with a paid retirement for Army officers who reached thirty years of service, enacted by Congress in 1870, and later the Navy's mandatory "plucking" (forced retirement) in the 1880s. In the Army, mandatory nondisability retirement could not be imposed on officers under the age of sixty-four. Despite these new retirement programs, there were no changes to seniority as the dominant factor in promotions until 1916, when the Navy adopted "promotion by selection" of impartial central boards. The use of selection as a policy was denounced as "scoundrelism" by many officers, reflecting a timeless concern about subjective bias and nepotism.

When Congress passed the Personnel Act of 1947, it formalized the battlefield flexibility of assigning and promoting officers based on the judgment of commanders rather than garrison seniority. That act formally gave the Army and newly created Air Force the power to promote by selection, although the selections were limited to cohorts of officers of the same age. The flip side of selection-based promotion was the mandatory retirement of officers nonselected for promotion. This was the "up-or-out" system pioneered by the Navy and extended to the Marine Corps by an act of Congress of 1925. The principle was limited to senior officers who failed to make flag rank, but it has crept down the ranks over the decades. In the aftermath of World War II, General Dwight Eisenhower testified before Congress that lockstep promotions until the grade of general officer were a serious problem.

Unfortunately, the up-or-out remedy of 1947 became a uniform straitjacket across all of the services in 1980. The enshrinement of a strict promotion timetable in the Defense Officer Personnel Management Act (DOPMA) of 1980 pushes all officers on active duty through the same career track and pressures nearly all to retire at their moment of peak productivity. Bernard Rostker and Rand colleagues summarized the act's effect in a 1993 report, which is equally accurate today:

> The "up" portion of the "up-or-out" system provides that, in general, officers move through the system in "cohorts" originally determined by the year of commissioning, and compete for promotion to the next higher grade against other members of the group at set years-of-service (YOS) points. The "out" portion of the "up-or-out" system provides that "officers twice passed over for promotion, after a certain number of years, depending upon their particular grade, are to be separated from active service, and if eligible retired."[104]

The great irony of the effort to cure the armed forces of its seniority problem is that "selection" became so formalized that seniority has been cemented into a lawfully mandatory rigid promotion timeline. In 2016, older officers are almost always higher ranked than younger ones, just as in 1866. The only difference is that the mass of officers are constantly churning up the ranks instead of stagnant in rank.

The standard retirement at twenty years of service and mandatory retirement at thirty years of service have a perverse incentive that overwhelms the small benefit of removing a minority of stagnant senior officers. The reality is that almost nobody has enough time at flag rank to become stagnant. They are competent, but a longer tenure would allow all flag officers to become far more proficient. Under TVF, the mandatory retirement age would be pushed back to forty years of service right away, and possibly later per service secretary discretion.

Pushing back the mandatory retirement age to forty years of service, and also changing retirement programs to eliminate the incentive to separate at twenty years for highly productive mid-rank service members, would sharply reduce the number of annual accessions.

Increasing average time in service by half would reduce the number of accessions by a third, reduce training costs by a third, and potentially increase the operations-to-training ratio as well. The military would get more productive human capital at a lower cost.

TABLE 5.1.
Time in Grade before Promotion

Rank	Required Time in Grade (years) All Svcs.	Average Time before Promotion (years)			
		Army	Navy	Marines	Air Force
(O-10)		32+	33+	30+	33+
(O-9)		33	31	35	31
(O-8)	1	30	30	30	28.5
(O-7)	1	27	29	27	25
(O-6)	1	22	21	22	20
(O-5)	3	16.5	15	16	15
(O-4)	3	10.5	9	10	9
(O-3)	3	4	4	4	3.4
(O-2)	2	2	2	2	2
(E-9)	3	22.6	22	22	23.6
(E-8)	1.6	18	18	17	20
(E-7)	2–3	12	14	12.8	16
(E-6)	1–3	8.2	8.8	8.8	11
(E-5)	6 months	4	4	4.6	5
(E-4)	2	2	2	3	2.5
(E-3)	6 months	1	1	1	1
(E-2)	6 months	9 months	8 months	5 months	6 months
(E-1)	6 months	6 months	6 months	6 months	6 months

Source: US Code § 619 and Institute of Land Warfare

Promotions and Timetables

The armed forces handle promotions in a way that must seem peculiar compared to other organizations. Promotion up to the rank of O-3 is largely automatic. Indeed, the promotion timelines are so rigid that the career trajectories of most officers look identical to most outsiders. This is driven by DOPMA's promotion timetable, the practical effect of which is shown in table 5.1.

DOPMA's "up-or-out" principle is so rigid that every branch of the armed forces promotes officers on the exact same timeline for the first ten or more years of service, and roughly the same for the second decade. The law allows service secretaries to extend *but not reduce* time-in-grade requirements for ranks O-3 and above.[105] It does this in order to make sure that officers get at least two opportunities for promotion board consideration.

In most organizations, an individual who is hired to fill a job is simultaneously promoted to the rank affiliated with the job. Because the military long ago cleaved the two, the complexity of conducting promotions followed by assignments has few outside comparisons. Getting a promotion does not mean you are getting a new job, and vice versa. Rarely does an officer's change in rank coincide with a new role. Indeed, pinning on a higher rank usually occurs while in one's current job.

The carefully orchestrated three-phase process is meant to maximize a theoretical fairness among all officers during every step while maximizing the needs of the military. Promotions occur first. Screening for job types (including command roles) comes second. Job-matching comes third. In retrospect, the actual "promotion" in rank really serves as a necessary qualifying step for future roles.

Officially, this is how promotions work for officers below flag rank.[106] A selection board committee convenes at the request of a service secretary, with the president's approval.[107] After approval, the secretary will compile a list of active-duty officers eligible for promotion to the next grade in each competitive category. The board's objective is to screen all officers on the list and rank-order them from top to bottom.

The secretary of defense determines what information may be provided to the selection board. To begin, the board is given a list of the needs of the force: particular skills, the minimum and maximum officers needed for competitive categories, and whether any officers have the needed skill sets.

Board members are given a copy of the officer's official military personnel record. A military personnel record includes very basic information, such as date and type of enlistment/appointment, duty stations and assignments, training, qualifications, performance, awards and decorations received, disciplinary actions, insurance, emergency data, and administrative remarks. Officers can submit comments to the board regarding the information. In addition, officers eligible for promotion are allowed to write a letter directly to the selection committee regarding their promotion. The letter must be under ten pages and meet the requirements of the selection committee. If the letter (or its attachments) do not meet the requirements, then it will not be shown to the board. Surprisingly, attachments that accompany the letter may not include recommendations, previous assignments, or education.[108]

The promotion board process famously allows each board member to view each individual "packet" for one minute and make an independent score. Scores are tallied and the list is generated. The final rank-ordered list is secret. The names of those selected for promotion are published in a list by order of seniority, and promotions are made "in the order in which the names of officers appear on the promotion list" using pin-on dates.[109]

In the Air Force, promotion to O-4 (major) slightly varies the timing such that some sliver of a year group (named after the year of commissioning) makes major a year before the main cohort, known as "below the zone." Half the year group is promoted a year later "in the zone," and then a distinct minority are promoted "above the zone." During 2014, for example, 86 percent of USAF officers in the 2005 year group were captains compared to only 30 percent of the older 2004 year group. This pattern reveals the annual bifurcation during which half of a year group is promoted simultaneously after ten years in uniform.

Naturally, the few officers not selected for promotion in the zone are known by all to be of a lesser class. Oddly, the list itself is kept secret, so no one knows the relative merit among those who are not promoted. Actual promotions are given out in order of "pin-on" dates, and the top 20 percent of selected individuals are designated as "school selects," but the handful of officers who just miss the cut are indistinguishable from the others (and ignorant of their relative status). The rank-ordering is valuable information, yet it is not utilized ever again—not by future commanders, future promotion boards, or the individuals in question. Indeed, a tremendous amount of valuable personnel information like this is expunged.

The laws governing promotions, notably DOPMA constraints, should be revised to allow service flexibility so that the chief of staff of the Army, commandant of the Marine Corps, chief of Naval Operations, and chief of staff of the Air Force can establish promotion rules that are best for their men and women. Even if the Army prefers to maintain the rigid timeline, the Navy would be allowed to loosen its up-or-out timeline, while the Air Force would be able to end the use of year group promotion zones entirely.

Mandatory promotion zones hinder the optimization of job-matching and specialization. The universal timetable requires the force shape of each service to look just like the others. However, it would not be useful to replace the DOPMA mandate with an alternative timetable mandate. Rather, each service should be allowed the flexibility to regularly reform its force shape and promotion timetable. That means the commandant of the Marine Corps could even continue using the DOPMA promotion zones.

If Congress does not amend DOPMA's timelines, it should at a minimum loosen the rigidity of the promotion zones by offering service chiefs flexibility on the issue. Each service should have expansive authority to use below-the-zone promotions for up to 40 percent of its officers in each cohort (double the maximum allowed of 20 percent).

The Air Force and, to a lesser extent, the Army use below-the-zone (BTZ) promotions in ways the Navy and Marines do not. The use of BTZ offers advantages in identifying strong, young leaders for promotion, but it has downsides in light of the other rigidities. One disadvantage is the "halo effect" that separates BTZ officers from others by offering better assignments, greater opportunities, and a continuation of special treatment, all based on the first early promotion. It's not that the BTZ process itself is flawed, rather that in-the-zone officers feel (and often are) permanently behind and are unable to catch up. This leaves little room for late bloomers or those circumstantially noncompetitive due to assignment timing. Oddly, BTZ officers also often find themselves at a disadvantage during slating when they are the lowest ranked individual in a cohort selected for command, often getting the least desirable job on the slating list. While the BTZ concept is not broken, its flaws could be remedied by scrapping the year group structure entirely. In the more open TVF structure, the idea of year groups would no longer be a factor in promotions or assignments. But this is because DOPMA requires nonpromoted individuals to be forcibly retired, which TVF would eliminate. The only service members required to be forcibly retired in the TVF will be those who are unable to find an active-duty job.

In order to control for dominance of any one category of officers in promotions (e.g., pilots in the Air Force), the law allows each service to establish competitive categories. Officers in the same competitive category compete among themselves for promotion. In the Air Force, only a few categories (medical corps, chaplains, judge advocates) are used outside of the main line. In the Navy, competitive categories are established for intelligence, information warfare, and other more general functions as well as the unrestricted line. This flexibility is not as useful as it might seem—especially in light of the critical need for advanced cyber and acquisition management skills—because each category is bound by the same career straitjacket per DOPMA promotion timetables. Congress should act to give United States Cyber Command, currently headquartered at Fort Meade, Maryland, an exemption from DOPMA's career structure as a unique and critical workforce. The same exemption should be granted to active-duty personnel in the acquisition workforce, plus allowances for service chiefs to define different timetables for different competitive categories.

One of the more revolutionary ideas that could improve talent management in the armed forces is to merge the assignment and promotion processes into one. This would replace the existing multistep process of promotions followed by assignments. The standard promotion boards would become purely *qualification* boards—designating a (larger) group of officers to be eligible for higher-ranked assignments and roles. There would be no pin-on dates for field-grade positions. Rather, the pin-on would occur simultaneously when a service member begins a job at a higher rank. Similarly, each service should have the power to make brevet promotions, a common practice in the Army dating back to 1776 that allows individuals to wear the rank affiliated with their current assignment and then return to their permanent rank afterward. While such a step is recommended, it is also unlikely in the short term, and it is not necessary to implement any of the other recommendations either.

Aiming High: Case Study in Choosing Commanders in Today's Air Force

Every autumn, the Air Force prepares a list of officers selected for squadron command. While taking command of a squadron is a momentous achievement in its own right, it is also the vital stepping-stone to future leadership roles. Management of the command selection process is similar in many ways to the promotion process, driven by a board of carefully chosen senior officers from all around the world tasked with using the fairest means to first create a list of individuals who are qualified to take command. Next is a slating process that selects from that list those individuals who are actually given specific jobs. The 2015 results were published in the following Air Force Times article:

> The Air Force has picked 514 officers for a variety of squadron command posts in 2016, the Air Force Personnel Center said Tuesday. The officers—who range in rank from captains to lieutenant colonels—will command logistics, support, materiel leader, air base, training, recruiting, test or medical squadrons, AFPC said. Wing hiring officials, major command functional managers and AFPC assignment teams matched officers from a list of potential candidates—which was released in September—to projected command positions. The assignment teams will have to coordinate with the units that are losing and gaining the new squadron commanders to decide what date the officers will report to their new assignments. Officials chose 17 captains, 153 majors and major-selects, and 344 lieutenant colonels and lieutenant colonel-selects. The development teams to choose squadron commander candidates began considering officers in May. They ended up with a list of more than 980 candidates from 29 career fields, the Air Force announced in October.[110]

Each major command (MAJCOM) manages its own group. For example, Phoenix Eagle is the name that Air Mobility Command uses for its command screening process, nominating a different group of mid-rank officers than those nominated by the other nine MAJCOMS. The process begins with officers of a certain rank submitting an official form letter to their MAJCOM volunteering to be candidates for command, a letter that must include language stating their acceptance of *any possible assignment*. Volunteering cannot be limited to certain types of units or locations, nor can any position be ruled out. Officers who are selected but subsequently turn down their assignments suffer negative career consequences, including possible forced retirement without pension. On the other hand, a qualified officer who does not "volunteer" for command is sending a dangerous signal that all but ends one's career.

Compared to the other services, the USAF process for selecting these commanders is highly flexible, with greatly decentralized authority over individual selections. Each MAJCOM has a development team that scores all of the individuals under consideration (or, rather, scores their packets). Development teams meet two or three times per year. The actual selection is done by senior commanders—for example, a wing commander choosing the individual who will lead one of her squadrons.

For Air Mobility Command, as an example, AFPC might generate a list of 180 officers eligible for squadron command from a crop of 250 volunteers. Of the Phoenix Eagle candidates, roughly fifty will be assigned to command a squadron, but this process involves a surprising degree of decentralization. Senior officers can bid on candidate officers up for command, a process that is refereed by AFPC. Note that different specialties have widely different promotion rates; some specialties see only 20 percent of eligible candidates selected for command, whereas others are critically understaffed and see nearly 100 percent of candidates selected.

Restore Command Authority

A more advanced TVF recommendation is for service chiefs to restore authority to unit commanders over personnel functions, especially hiring. It may surprise some readers, but restored hiring authority can be implemented by any service without legislative action. The centralization of promotion boards and assignment centers occurred principally during the early 1960s under the direction of Secretary of Defense Robert McNamara.

The inefficiencies of central planning plague military job-matching. Critics of decentralized hiring echo concerns about promotion-by-selection made a century ago. What is to stop bias and favoritism? There are concerns about sexism, racism, or good-old-boy networks. The challenge is to design a process that maximizes the efficiencies of decentralized optimization while minimizing bias. The status quo aims to be so impartially centralized that it suffers from matching inefficiencies and outright ignorance of service member preferences. Interestingly, each of the armed forces has the legal power to decentralize command authority to whatever degree it deems best while retaining rules that will prevent bias.

To prevent biased hiring, TVF makes sure command authority is not absolute. It maintains a role for central boards and branch managers—by screening which individuals are nominated for assignment to commanders (at the rank of O-5 and above)—to exercise final authority over hiring. See figure 5.3 for a look at the basic idea of job-matching in the status quo, where the commander's function is idle, compared to the TVF, where it is active and vital. Personnel centers/commands will provide a slate of no fewer than three candidates for the unit to interview and choose for key roles. Commanders should have limited authority to directly hire, whereas most hires will be through the centrally provided slate of candidates. Many key developmental roles should still be directly assigned centrally—meaning that a single candidate shall be recommended by personnel centers in many instances (e.g., honoring follow-on assignment commitments)—but the unit commander should retain the right to veto a limited number of such assignments.

More local authority should also be granted to the labor supply side. That is, the TVF will give more career control to individuals in

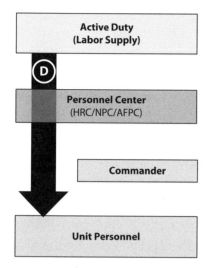
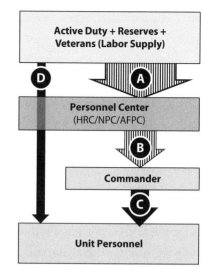

FIGURE 5.3. Job-Matching in the Military: Status Quo versus the Total Volunteer Force

A Individual service members who are eligible and qualified for rotational assignments APPLY to open positions listed in the online talent management system.

B Volunteers will BE SCREENED by managers at the Personnel Command/Center (PC) and three candidates will be recommended to Unit Commanders.

C COMMANDERS will interview candidates and make an offer to their top choice.

D Assignments are made DIRECTLY by managers at PC. This includes promised follow-on assignments and key development positions deemed essential for select individuals.

Source: Author

choosing assignments. To repeat, these reforms are not being recommended for all ranks. Rather, the TVF recommends implementation for mid-career ranks.

The TVF process outlined in figure 5.3 would differ from the status quo by allowing individuals to decline assignments. Instead, it would empower individuals to apply for open jobs. To thicken the labor supply, a service could be allowed (but not required) to make some jobs open to reservists and veterans. To be sure, individuals should only be able to apply for jobs for which they are qualified. This new approach to job-matching raises a question about the capacity of the existing structure to accommodate a flood of applications. How would commanders have time to sort through all of the applicants?

This important question is not so daunting as it might seem, especially given the unlimited number of applications to jobs outside of the active force. A second question is how to prevent favoritism and bias in the hiring process. The answer to both questions sketched in figure 5.3 is to involve the personnel centers in winnowing the applicant list to a small number of nominees for a unit commander's consideration.

A workable system could accommodate the technical supply of labor in many ways. Individuals could be allowed to apply for up to five (or ten, or fifteen) open jobs at a time. Alternatively, individuals could designate a handful of jobs in preferential order while also expressing willingness to accept dozens more. But couldn't service members (and reservists and veterans) still flood the computational capacity of detailers who will be tasked with screening the applications? That is unlikely. But a simple remedy is to require individuals to prioritize their job applications using a modified "dream sheet." One workable TVF dream sheet process involves a list the job-seeker maintains with three "top" choices, up to seven "preferred" jobs, followed by an unlimited number of "acceptable" jobs. The list could be constantly updated.

An avenue that should be encouraged is to allow individuals to query units and HR officers about open and forthcoming jobs, including potential job openings in the dream sheet. Lastly, the TVF should allow individuals to opt out of the promotion cycle in order to extend and specialize in their current roles. Likewise, individuals should be allowed to apply for open jobs at their current rank, rather than be forced to move "up" during every change of station. This would allow, for example, aviators to stay in the cockpit, cyber warriors to remain in critically understaffed positions, and combat commanders to extend their tours during wartime operations in which continuity is vital for mission success. More flexibility and cost savings will be achieved if individuals can apply "down" to lower-ranked jobs if that is their preference.

Turning back to the labor demand side, the TVF will allow commanders to make fast dismissals and emphasize quick replacement. Currently, commanders technically have the authority to remove a subordinate whose performance is subpar through a lengthy, punitive, and bureaucratic process that is rarely used due to its cumbersome nature. The services should empower commanders with a flexible array of options to include for-cause firing—but the critical missing piece is simply to allow a no-fault dismissal, an action that would not reflect negatively on the service member.

Dismissals and firings of individuals currently leave a hole in the unit—the unfilled billet problem—which penalizes commanders. This tends to keep poor job matches in place, with the appearance of a smoothly functioning organization, but it is rooted in a perverse incentive to maintain inefficient matches. The dismissal process must be fixed to allow quick replacement.

Turning lastly to the question of capacity: would a talent market described here burden the time of a unit commander and staff? Perhaps. The truth is that most corporate managers spend a much greater percentage of their time on personnel management as opposed to operations than military commanders. Indeed, corporate leaders often marvel at how disengaged military leaders are at coordinating best-fit talent. While the TVF may require more up-front costs on the commander's time in coordinating talent, it offers unrealized gains in readiness as well as the commander's time because it will avoid current issues with teams that are poor fits with some unit members.

Timing and the Talent Marketplace

A fundamental problem with the existing assignment processes is that it is not built for real-time job placement, which makes nimble replacement moot. The decentralized TVF assignment process described earlier could be implemented regardless of a real-time assignment norm (and with no need for legislative changes), but it would benefit from a continuous job-matching process as opposed to the current discrete process conducted two or three cycles per year.

A discrete job-matching process tries to optimize n faces against n places all at once. There are many downsides to this status quo, the biggest being that the real world does not work that way. Often, the optimal job for an individual (in terms of desire or key career progression) is not available during the current job cycle. Square pegs are placed into round holes all the time. The TVF would instead let a square peg wait until the better fit becomes available. In other words, the TVF job market will be continuously open. Any service member could log onto the online TVF talent marketplace in order to see the current list of open jobs. Once a job is filled—meaning the assignments officer has forwarded three nominees to the unit commander, whose staff has reviewed and made a selection, and orders are issued and accepted—then the job listing would be immediately removed from the marketplace.

The notion of a talent marketplace has already been piloted in the US Army's proof-of-concept project known as Green Pages from August 2010 to August 2012. Designed to study and potentially resolve many of the dynamics discussed in this book, Green Pages was limited to officers in the Engineer Regiment. In two years, nineteen discrete iterations made 748 actual assignments. Ten iterations were for captains, seven for majors, and two for lieutenant colonels.

The pilot project was a success, but not a total success. First, it showed that a marketplace could be used to enhance rather than displace the existing role for branch managers. Second, Green Pages engaged individual officers in order to obtain a much richer profile of the skills, education, and talent already in existence. By offering a new incentive to build an attractive and accurate profile in the system, individual officers themselves entered a tremendous amount of data that was unknown by the older top-fed central system—languages spoken, countries visited, engineer certifications completed, and so on. In fact, half of the participants found at least two significant errors in the top-fed data. More to the point, the Army had never been able to incentivize profile data participation on the wide scale that Green Pages did. Third, officers expressed greater satisfaction with the outcomes (notably because participants were 34 percent more likely to get their top preference). Fourth, officers and gaining units *changed* their match preferences during the market phase of each Green Pages iteration. That is, each side learned more about the other—particularly when the other side expressed interest—and adjusted its preference rankings, allowing branch managers to make more optimal matches.

Efficient job-matching requires an information system that makes available jobs (requirements) visible with details about location, unit, role, commander, and more. Unfortunately, the downside of Green Pages is that units were not as active in providing details about open jobs, even though they did participate in the market operation. The after-action report on Green Pages described efforts to fix this aspect of weak demand-side information, which ultimately involved essentially haranguing unit strength managers to participate. The problem was chalked up to cultural and training biases, acknowledging that a few unit commanders did not grasp the concept or need. The fact is that many units had one or two engineers out of many other officers. The commander, regardless of participating in Green Pages, could be confident that the engineering position would be filled. In other words, there was no negative consequence for ignoring the market-place—*and there never will be a demand-side incentive in a discrete assignments process.* Not so in a continuous job-matching marketplace.

Let's step back and imagine a branch manager matching *n* infantry majors against *n* infantry requirements during a given cycle. Let's imagine *n* = 100, for the sake of argument. Currently, the manager will try to optimize the match, generating 100 sets of orders on the final day of the process. An alternative would be to cream the top matches (that is, the twenty or so individuals whose top preferences and qualifications are met by being the top preference of twenty or so units) and then re-run the match, allowing individuals and units to see the remaining eighty available faces/places. Preferences would be reordered. If this process is iterated ($n = 65$, then $n = 50$, and so on until $n =$ the last 10), then a different hundred matches will be made because more detailed preferences are revealed.

An even better process would allow a long market phase during which participants could see the preferences of the other side (demand sees supply preferences, and vice versa) and then update their own. Interviews could be conducted. Officers could e-mail questions and get answers. Each iteration of Green Pages had a market phase of two to six weeks allowing exactly this kind of back and forth, but it had a single postmarket assignment phase in which all n matches were finalized.

Currently, the US Navy is developing a web-based talent management platform called Talent Link, which builds on the concepts piloted in Green Pages. It envisions discrete job-matching cycles with three periods—discovery, evaluation, and slating—which involve profile-building, market interactions to finalize preference lists, and job offers, respectively. The Naval Personnel Command aims to run a pilot program for multiple officer communities, including surface warfare officers.

Recognizing that a continuous talent market would involve revolutionary change, the following questions arise: How does the system resolve individuals who neglect to submit a preference list or overstay their current jobs? How does the system anticipate job openings if individuals are able to overstay in their current jobs? Can individuals job-hop as often as they want? And if not, what role do losing commanders have in retaining their unit members? What happens to units if a commander selects a nominee, but the nominee rejects the match? What if this happens multiple times? What if the sweetheart jobs are only open for a few hours before being filled? All of these dilemmas will be resolved relatively smoothly in practice, but resolution may require jettisoning the discrete cycles in place now.

The TVF Job-Matching Process

The following twelve guidelines describe essential aspects of the TVF's continuous talent marketplace:

1. **Requirements.** Requirements posted in the marketplace are denoted as (1) open jobs, (2) jobs with nominees, (3) filled jobs pending acceptance, or (4) filled jobs with orders. An open job will be posted by a gaining unit and must include details about the associated rank (minimum and maximum), specialty code, location, unit commander, roles and responsibilities, and required skills and certifications. Another detail that must be listed is the pay, including all associated allowances and bonuses. Note that the assignment incentive pay will be included—a program described in the next chapter on compensation—which automatically increases for long-unfilled positions according to a formula.

2. **Maximum and minimum tenure.** Each requirement will also have a designated minimum tenure, ideal tenure, and maximum tenure. For example, a job listing for a captain in an armor battalion might list a minimum tenure of eighteen months, an optimal of twenty-four months, and a maximum of thirty-six months, meaning that the assigned individual would not be permitted to leave for another job sooner than eighteen months. Individuals will be encouraged to begin searching and applying for jobs six months before their minimum tenure. They will be alerted nine months before their maximum tenure. Maximum tenure for some jobs can be extended,

with approval of the unit commander, central manager, and service members, three months prior to the date.

3. **Key jobs.** Some jobs are key for developmental purposes, but many are not. For example, battalion command is a critical career experience for mid-level Army officers aspiring to high rank. For non-key jobs, a very long or unlimited maximum tenure should be the norm to enable longer tour lengths that will reduce rotation costs and burdens. Congress has already encouraged this action in recent hearings and reports. The TVF marketplace will allow individuals to extend their current tour lengths for one to two years, indefinitely, with approval from the unit commander.

4. **Open window.** Any job posted in the marketplace must remain open for a minimum number of days (let's say twenty) before the central manager is allowed to formally select at least three nominees. Urgent postings and repostings approved by the central manager can lower the minimum posting time, but never less than seventy-two hours. This ensures that service members have a reasonable opportunity to review all jobs.

5. **Application.** Individuals searching for open jobs will maintain a dream sheet (visible by commanders and central managers) of their preferences: top three, up to seven preferred, as well as an unlimited number of acceptable jobs. Individuals are permitted to call a unit point of contact to potentially learn more. Veterans on active and inactive reserve status may also apply for jobs in the same fashion, but only for jobs that each service designates as open for lateral reentry.

6. **Personnel managers select nominees.** Personnel center managers (i.e., detailers) will screen applicants and select a small number of nominees for each open job. Three nominees seems to be an optimal number. Their names and files will be sent to the commander of the gaining unit.

7. **Notification of nominees.** The system will automatically notify the nominees of their selection via e-mail and will notify other applicants that (nonidentified) nominees have been selected. The notification allows nonselected individuals to update their job preferences.

8. **Commander's hiring process.** The unit commander is given a set time to make a selection in the form of a prioritized list. The commander has the right to review any and all available information about the nominees, to include personal interviews, reference checks, and online profiles. The commander has the authority to hire, offering one of the nominees the job. The commander also has the right to reject all of the nominees and request the job be reposted. The commander may not reject all of the second slate of nominees.

9. **Acceptance.** Selected nominees may turn down one assignment order per year. They bear the risk of potentially being put on reserve status.

10. **Priority cases.** The central manager can nominate a single nominee, without any open job posting of the requirement, for certain priority individuals and cases. The most common will be promised follow-on assignments and key developmental assignments for select individuals. Each service can designate the maximum percentage of special cases per unit. Note that the commander has the right to reject the first nominee, even in special cases, though this can be overridden by a general officer.

11. **Hitting maximum tenure.** At some time before the maximum tenure date, individuals may be involuntarily assigned to a new job by central managers. If the maximum tenure date is reached, and the individual has not been selected for and accepted assignment to a new job, he or she will be placed on reserve status on half pay and full allowances for a set time before pay is suspended. At all times, including paid and unpaid reserve status, individuals

can apply for open jobs on the talent marketplace. After twelve months on unpaid reserve status, the individual will be considered honorably discharged.

12. **Transition time with orders.** Once orders are issued to complete a talent-job match, ample transition time should be the norm. Losing units in particular must be given time to post an open job requirement and hire a replacement. Each service can design the transition periods as needed for different types of jobs.

One of the realities of the existing manpower system is that in some services there are many more unfilled requirements than available personnel. Thousands of requirements are left empty for years, and the reason is that it costs the armed forces nothing to do so. People are a cost; empty seats are not. Eliminating low-priority requirements is, however, costly in terms of time and hassle. Why bother? What manpower managers should do, and must do for a TVF transition, is to perform a thorough prioritization of existing requirements. As a default, a requirement should be considered standard-priority. Managers would have to proactively designate others as higher priorities (key experience, command track, etc.) and, as part of the review, establish the tenure points for these higher-priority jobs. Only priority jobs would qualify for pays above the basic rate (explained more fully in the next chapter).

Does the TVF require a new online marketplace? What is described above involves an online market, but many TVF reforms could be implemented without it. For example, steps (6) and (8) could be implemented today. Personnel centers should screen and nominate three individuals for unit commanders, who will make the hire. No new technology is required for this core process. What technology can do is enhance the efficiency of the process. Matching algorithms can help service members screen all available jobs in order to make sense of what otherwise could be information overload. A web-based talent-management database could expand the amount of information—as Green Pages did for Army engineers—about faces and spaces. Technology can empower service members to build more nuanced dream sheets that express their preferences. Yet as military members know too well, development of technology is slow and often yields worse outcomes due to poor implementation. That is why TVF emphasizes process reforms first, technology second. Establishing command authority is far more critical to reform than building a new master database.

Will the TVF strain the time capacity of commanders, especially if it lacks new technology? Yes and no. There is no question that giving the authority to hire to commanders rather than central planners also burdens commanders with a newfound responsibility. It is not clear this will be more than a light and productive burden, and it promises to add profoundly to unit readiness and morale, an investment of upfront time that will yield dividends in freeing up command time on operations. To avoid this capacity constraint from becoming a logjam that breaks the assignment process, reform must include well-designed defaults. Specifically, a commander who neglects to make a nominee selection within a set timeframe defers the choice back to the personnel command. The services could pilot different default timeframes and refine them going forward. The biggest mistake of all is to assume the reformed job-matching must be perfect and rigid from day one. Flexible defaults will be essential during the years of transition.

CHAPTER SIX

Reforming Compensation

T he military's compensation structure creates perverse incentives that are at odds with voluntary service. Why does the Pentagon offer a retirement system that vests after twenty years? And why does the Pentagon allow that pension to be drawn at exactly the same moment instead of at age sixty-five? Is this pension structure not (1) using coercion to keep some employees in place longer than they otherwise would choose to remain and (2) creating an incentive for them to retire promptly, at the moment of peak proficiency?

Military service is motivated by selfless love of country, but not that alone. The military relies on other incentives, too, including pay, health care, retirement, training, education, camaraderie, travel, and adventure. Unfortunately, the military's complex compensation system is overly expensive and inefficient—two factors that are closely related. Culturally, some military members like to believe that *money* and *service* are at odds, a belief that allows them to neglect how crudely and poorly the Pentagon compensation structure is designed.

The economic incentive is and has always been part of Pentagon planning for manpower. Incentives matter. As economist Eric Hanushek wrote in 1977, "Many of the largest personnel problems [in the US military] are exacerbated, if not caused by, the incentive system." The main theme underlying the Total Volunteer Force (TVF) is that coercion is less efficient than volunteerism. And compensation reform is perhaps the easiest way for the Pentagon to improve talent incentives in place of coercion.

The stark reality is that military pay offers no monetary rewards for excellence. The armed forces spend more funds moving personnel around the globe than on performance rewards. More than 90 percent of personnel spending in 2016 went for base pay, compared to 3.9 percent on permanent change-of-station (PCS) moves and half of an percent on "incentive and special pays." That phrase is a misnomer, as incentive pays do not reward performance; rather, they are designed to compensate career specialization and provide broadly available retention bonuses.

Three reforms to military compensation will increase efficiency and morale while reducing gross expenses. The first is to adjust the standard military pension in a way that gives more options to service members and establishes a trajectory to eliminate twenty-year cliff vesting. The second reform is an adjustment to the base pay schedule that ends the use of tenure and replaces it with role and responsibility pays. The third reform is to utilize existing incentive pays to make the new job-matching program work efficiently—a way to reward individuals who volunteer for tougher assignments. We will lump the latter two into a single section below, but first we will review the problems identified in chapter 1.

Leader/Talent

The Leader/Talent results presented in chapter 1 include four elements concerning compensation (figure 6.1). Scores range from always true (+2) to always false (−2), and all four compensation elements were scored negatively by military respondents. The average score on compensation does not mean that military paychecks are low. Rather, it means that the compensation process is evaluated

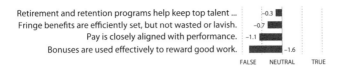

Retirement and retention programs help keep top talent ... –0.3
Fringe benefits are efficiently set, but not wasted or lavish. –0.7
Pay is closely aligned with performance. –1.1
Bonuses are used effectively to reward good work. –1.6

FALSE   NEUTRAL   TRUE

FIGURE 6.1. Leader/Talent Measures of Compensation Elements in the US Military
Source: Author's Leader/Talent survey

poorly regarding rewards for performance, good work, efficiency, and retention. The single lowest score goes to "bonuses are used effectively to reward good work" for the simple reason that the military has no compensation mechanism to reward or penalize performance. In addition, there is a huge gap between military and civilian organizations on two of the four elements: "pay is closely aligned with performance" and "bonuses are used effectively to reward good work."[111]

Changing compensation is not necessary to make effective reforms to other personnel processes such as promotions and job-matching, but it will improve them. And even by itself, a modernized compensation system can enhance talent management and morale.

Rather than suggest merit-based pay, an approach that has proved controversial in government practice, the TVF instead recommends shifting base pay away from seniority and toward the specific role a service member fills. In essence, paying individuals for the exact job they perform is the best way to reward merit. The more meritorious individuals will take on harder assignments, in more remote locations, involving greater responsibility—and consequently receive higher pay to reflect those choices.

Unheeded Compensation Commissions

Every four years, the Pentagon conducts a high-level review of compensation known as the Quadrennial Review of Military Compensation, or QRMC. President Obama canceled the one due in 2016 on the grounds it was "not required" in light of the congressionally created Military Compensation and Retirement Modernization Commission, or MCRMC.[112] That commission published its recommendations in early 2015, and its proposal to change the military retirement system was adopted into law that year. It was an impressive achievement on one hand, given that dozens of blue-ribbon panels had failed in similar endeavors over the decades, but the change itself was not as monumental as many observers seem to think. The retirement pension for new active-duty members will be reduced by one-fifth (starting in 2018) in order to develop a new defined-contribution savings asset, but four-fifths will remain, unaltered in vesting at twenty years and payment immediately upon separation. The incentives have not been altered in any fundamental way.

The first QRMC was convened in 1965, and as Bernard Rostker noted in his history of the All-Volunteer Force (AVF), subsequent QRMCs often splintered over varying controversies. The implicit question about the purpose of compensation was often debated, despite the seemingly obvious answer—settled thanks to research by Beth Asch, John Warner, and (separately) Curtis Gilroy—to "attract, retain, motivate, and separate personnel."[113]

The single most important compensation commission assembled in the past century was the Gates Commission, created at the presidential level in March 1969 at the beginning of the Nixon administration in order to evaluate the question of whether to replace the draft with a system of manning the military entirely with volunteers. The commission was led by Thomas S. Gates, who had served as secretary of defense under Eisenhower from 1959 to 1961 and who was initially skeptical of the proposal. Reporting just under a year after its formation, the members made a shockingly unanimous recommendation to end the use of conscription. What is neglected by history is that the Gates Commission also recommended a set of comprehensive compensation reforms it said were vital for making the AVF operate effectively. Those reforms, aside from increasing base pay to attract volunteers during accessions, were "unheeded in the years that followed," according to Todd Harrison's 2012 study[114] of what kinds of compensation are actually valued by active-duty troops: "As a result, the fundamental structure of military compensation remains largely unchanged despite the transition to an all-volunteer force."

A summary of the compensation recommendations made by the Gates Commission in 1970 included:

- Increase overall base pay.
- Add supplemental pay for multiple levels of hazardous duty as a "matter of equity as well as to provide compensation flexibility."
- Transition to a salary system, meant to replace the inefficient mix of in-kind allowances (housing, subsistence, facilities) and deferred compensation that has "little value for a new recruit."
- Eliminate periodic reenlistments for specific time periods with an open-term work contract.
- Allow greater choice of occupation.
- Allow lateral hiring of skilled civilian personnel into higher ranks.
- Reform military retirement, including lowering the vesting eligibility from twenty years to five years and raising the receipt of pension closer to normal retirement age. The rationale was that individuals "cannot afford to separate from the service" beyond their tenth year of service and the structure "induces many individuals to retire as soon as they are eligible."

There have been major negative consequences of ignoring the Gates Commission. The first is galloping inflation in the net cost of military personnel since 1973, even as the size of the force on active duty has steadily declined. The second is a hugely inefficient bubble of

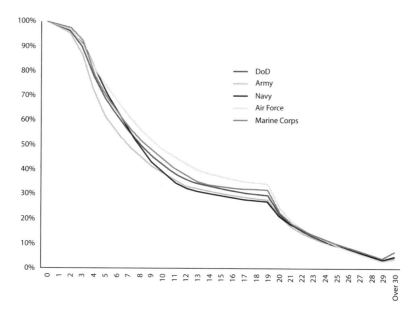

FIGURE 6.2. Continuation Rates of Officers by Years of Service
Source: Military Compensation and Retirement Modernization Commission, Final Report (2015)

active-duty manpower around the twentieth year of service (YOS), coinciding exactly with the twenty-year defined benefit (DB) cliff pension, unchanged for a century, which establishes perverse and co-ercive work incentives that are at odds with voluntary service.

Consider figure 6.2 from the MCRMC that shows the continua-tion rate of officers in the armed forces by years of service. There is a distinct twenty-year notch. Very few individuals leave the service after the twelfth YOS, and an alarming number quit promptly at the twentieth YOS. This is rational behavior given the incentives, but the incentives are wasteful.

Pentagon data on retirements by years of service also show a bubble at the twenty-year mark (table 6.1). There were 451,522 total non-disability military Army retirees who received retired pay in Septem-ber 2011. More than half of the retirees left during the twentieth year of service (234,462), and another 20,000 left through early retire-

TABLE 6.1.
Military Retirement Bubble at 20 Years of Service[115]

Service	Retirees	(20 YOS)	Early Retirees
Army	451,522	51.9%	4.4%
Navy	392,816	51.2%	3.6%
USAF	541,417	47.9%	3.0%
USMC	85,464	47.4%	5.4%

Source: Author's calculations from DOD actuary data

ment. In sum, 56 percent of Army retirees left active-duty Army service at the first retirement opportunity. The same pattern holds in the other services.

The retirement bubble at twenty years is followed by a smaller bubble at the thirty-year mark, which is when the multiplier of years in service is maximized. Serving beyond thirty years does not increase the multiplier. Across the Department of Defense (DOD), there were 38,600 retirees with thirty YOS compared to 6,034 retirees with thirty-one YOS and roughly 10,000 retirees for all years beyond that combined. These two bubbles are clear evidence that the vesting cliff and multiplier cliff create strong incentives to stop working on active duty.

As for costs, the twenty-year defined benefit pension is growing exponentially more expensive. The Defense Business Board reported in 2011 that annual outlays exceeded $50 billion in 2010 and will more than double before today's lieutenants become generals. Liabilities of the program are $1.3 trillion (roughly 10 percent of US GDP) and will rise to $2.6 trillion in 2035. This is "unsustainable," say Pentagon actuaries and countless scholarly studies.

There are other cost drivers. The tenth QRMC, published in early 2008, reported that net costs of pay and benefits were respectively $10,600 and $17,800 higher for enlisted members and officers than for their civilian peers. Mark Cancian described the updated cost pressures and the widespread concern that they are harmful to national security in a 2015 article in *Joint Forces Quarterly*:

> Since 2001, pay per Active-duty Servicemember has grown over 80 percent in 2001 dollars (about 50 percent in constant dollars). Military pay has increased 40 percent more than civilian pay since 2000, and enlisted Servicemembers are now paid more than 90 percent of their civilian counterparts with comparable education and experience (officers earn more than 83 percent of their civilian counterparts).[116]

So? Isn't higher pay the market premium for serving in uniform, given the danger and other unique burdens involved? Yes, but as Cancian noted, when active-duty military personnel get more expensive, DOD adapts by using more reservists, contractors, and capital.

Recall the problem identified by Hanushek, cited in chapter 3, of implied contracts with absolute career control by the military, which must compensate accordingly. Hanushek also decried the undifferentiated nature of these implied contracts—paying soldiers the same whether or not they are based in the least desirable locales, paying pilots a bonus whether or not they actually fly, and so on. We should extend the Hanushek problem further by analyzing the additional cost necessitated by a closed labor pool. The nature of a centrally managed labor market involves trying to perfectly match the number (and specialty) of requirements with people, i.e., spaces with faces. The challenge to precisely match numbers puts the personnel commands on a knife-edge with costs on each side: overmanning and shortfalls. In fact, the eleventh QRMC in 2012 included a chapter that examined four critical fields—remotely piloted vehicles, special ops, mental health, and linguists—and even calculated the "requirements ratio," which was significantly below 1.0 for these four specialties.

The military's closed labor pool is a profound cost driver. Hanushek observed in his 1977 paper just four years after the AVF's adoption that the closed nature of the military labor pool and other unreformed compensation policies would cause higher cost burdens than expected. Economically, the armed forces have engineered a highly inelastic labor supply. If the labor supply curve shifts inward (the incessant threat of the one individual most likely to quit), the military can only maintain equilibrium quantity of labor by escalating compensation. The standard economic toolkit offers only one solution to this problem: thicken the labor supply in order to make it more elastic. This is the effect of opening the labor pool to later entry, shown in table 6.2, leading to pay at, rather than above, equilibrium.

As an analogy, imagine trying to hire a plumber with five years of experience or more. If conscription were an option, this uniformed plumber could be paid below-market wages. But in a closed market where plumbers need to be accessed untrained five years prior and retained in a system of widely equal pay scales based on retaining the marginal individual, he would necessarily be paid above-market wages. To alleviate that pressure, the market must be opened to lateral entry at the mid-rank specialties. The closed system in place now is a one-way dynamic in which individuals have the ability to move out of but never into the ranks, albeit with rare exceptions. A closed labor market puts upward pressure on compensation.

TABLE 6.2.
Compensation Pressures under Different Labor Market Institutions

	Conscript	Volunteer
Closed to Lateral Entry	Pay below equilibrium Pre-1973	Pay above equilibrium 1973–present
Open to Lateral Entry		Pay at equilibrium (TVF)

Source: Author

In sum, there are two institutional distortions in the Pentagon's compensation system that artificially drive up costs. Worse, as Harrison has shown, a great deal of military compensation is valued far less by young troops than it costs the services. The Total Volunteer Force would remedy both with two general reforms. First, transform base pay by shifting from tenure to role and responsibility pays. In addition, add assignment incentive flexibility to help fill difficult jobs. Second, transform the retirement system. The MCRMC-inspired reforms of 2015 did not go nearly far enough. A detailed case for each major reform is presented next.

Transform Base Pay from Tenure to Role and Responsibility

Concerns about military compensation often focus on the escalation in pay and benefits. Total compensation per service member has nearly doubled since 2001, while the size of the active force is actually lower. The Government Accountability Office noted that service member earnings are equivalent to the 80th percentile of comparable civilian compensation as of 2008, up from the 70th percentile just four years earlier.[117] While escalating costs may be a problem, the more vexing issue is that the service chiefs have essentially no authority to shape and thereby control pay escalation. Rather, Congress determines annually what pay raise shall be applied to the one-size-fits-all pay table. Intelligent critics can complain that Congress should exercise more restraint, but there is an unavoidable incentive problem: military pay raises should not be a single lever managed by a legislative vote. In other words, why is there a universal pay table in the first place, instead of granting each service chief the authority to establish pay and allowances for their services within a fixed personnel budget? Flexibility is the solution to both the political economy problem that drives inflation and the readiness problem rooted in a rank structure constrained by the industrial-era hierarchy.

TABLE 6.3.
DOD Personnel Spending for FY 2016 (in $ millions)

	Total ($m)	Total (%)
Pay, Officers	33,374	29.5%
... Incentive Pays	488	0.4%
... Special Pays	1,189	1.1%
... Allowances	555	0.5%
Pay, Enlisted	68,655	60.7%
... Incentive Pays	245	0.2%
... Special Pays	1,730	1.5%
... Allowances	2,356	2.1%
PCS moves	4,460	3.9%
Total	113,054	

Source: Department of Defense FY 2017 President's Budget (M-1).

In this section, I will showcase the current pay table—the essential nature of which has been unchanged since the mid-1940s—and an alternative pay table that could improve flexibility, morale, and readiness, and even allow cost savings.

Active-duty pay and benefits total $33 billion for officers and $74 billion for enlistees, about half of each being base pay and the rest including retirement accruals, housing, subsistence (meals), social security taxes, and roughly 5 percent for special pay and allowances. Table 6.3 details a few of these items for each service.

Military *base pay* has been calculated on two dimensions since at least 1949: rank (pay grade) and cumulative time in service. The official pay table, which applies to all four armed forces, is organized in columns for time in service every two years (up to forty) and rows for pay grades (E-1 to O-10). Table 6.4 is an abridged version, but the reality is that most careers follow a narrow path along the diagonal. For example, there are no O-1s in uniform at the ten-year point, except for the rare individual with prior enlisted time, and there are certainly no O-5 and higher officers with less than two years of cumulative time in service.

Note that monetary rewards in the base pay table are greater for tenure than rank. For example, an O-1 typically is promoted to O-2 at the two-year point. Base pay rises $453 per month due to the higher rank, whereas the tenure step is worth $475. For O-2 to O-3, rank is worth $593, while time in service is worth nearly $800. The next promotion to O-4 at year twelve pays roughly $600, whereas an additional decade of service yields $1,200 more in monthly salary.

The use of tenure-base pay is intended as a crude proxy for level of responsibility—crude in the sense that a ten-year captain or sergeant is assumed to be 20 percent more responsible than a four-year captain or sergeant, and 10 percent less responsible than a sixteen-year person of equivalent rank. The use of tenure-base pay is an anachronism of a day when personalized algorithms that factor in occupation, location, merit, and detail-difficulty pays were impossible to imagine, let alone implement. Those days are gone. Why use crude proxies when the technology exists to fairly and impersonally designate pays for each role, while also fairly and impersonally adding bonus pays for critical roles that are difficult to fill?

Pay has always been supplemented by untaxed allowances for housing and subsistence, which are also unrelated to merit. These in-kind payments have roots in military tradition long before the nation was founded, but we know that the Navy and Marine Ration Chart of 1794 prescribed daily rations. For example, Monday rations were "1 pound of bread; 1 pound of pork; ½ pint of peas or beans; 4 ounces of cheese; and ½ pint of distilled spirits."[118] Most sailors were allotted one ration per day, specialists such as carpenters and boatswains received two, lieutenants three, and captains six. Today, common allowances include $530–$2,000 monthly for housing, more for

TABLE 6.4.
2016 US Military Base Pay Table ($)

Pay Grade	Under 2 yrs	Over 2 yrs	Over 4 yrs	Over 10 yrs	Over 16 yrs	Over 20 yrs
O-8	9,946	10,272	10,549	11,374	12,293	13,319
O-7	8,264	8,648	8,967	9,768	11,269	12,044
O-6	6,267	6,885	7,337	7,722	8,937	9,848
O-5	5,225	5,886	6,370	7,110	8,159	8,617
O-4	4,508	5,218	5,644	6,746	7,449	7,527
O-3	3,964	4,493	5,287	5,998	6,448	6,448
O-2	3,425	3,900	4,644	4,739	4,739	4,739
O-1	2,972	3,094	3,740	3,740	3,740	3,740
E-9				4,949	5,368	5,704
E-8				4,230	4,618	5,138
E-7	2,816	3,074	3,347	3,796	4,298	4,550
E-6	2,436	2,680	2,914	3,409	3,720	3,825
E-5	2,231	2,381	2,614	3,148	3,166	3,166
E-4	2,046	2,150	2,382	2,483	2,483	2,483
E-3	1,847	1,963	2,082	2,082	2,082	2,082
E-2	1,757	1,757	1,757	1,757	1,757	1,757
E-1	1,567					

Source: Defense Finance and Accounting Service

overseas housing, money for uniforms, and a $250 monthly payment for jobs that require family separation.

There are over sixty special and incentive pays, according to the Defense Finance and Accounting Service,[119] including long-standing pays for sea duty, flying duty, hazardous duty in combat, reenlistment, and other retention bonuses used for force-shaping. Special pays in 1949 were narrowly available, including sea and foreign duty pay for enlisted members of different ranks (ranging from $8 to $22.50) and aviation or submarine crew pay.

It must be understood that incentive pays are *not* designed to reward performance; rather, they are designed to compensate career specialization and skills retention. And this is what makes the history so instructive. Differential pay for occupational specialties is relatively rare today, with important exceptions: doctors, pilots, and especially Navy nuclear propulsion specialists (Hanushek said it "only occurred in extreme cases" in the 1970s), but it was common a century ago in the armed forces. Today, most troops of the same rank get roughly the same monthly paycheck, only because the US military devolved away from differential pay step by step during the twentieth century. In 1893, for example, monthly pay in the US Navy was highly differential across eighty enlisted ratings, from $9 (for apprentices, third class) to $70 (for machinists), whereas most sailors earned $24 to $35.[120]

In contrast, the use of tenure in pay calculations emerged in the late nineteenth century. The Navy pay chart for 1860 has five columns for "years on sea service" for officers, with the lowest, "Under 7" increasing by two-year increments up to the highest, "Over 13." Either an officer was promoted to the higher (and higher paid) rank or stayed in the senior lieutenant pay bracket. In 1870, this had been replaced by a binary tenure column for each rank, with a bonus after the five-year term *in that rank*.

Today, differential pay is utilized for some specialties, but not to the degree necessary for optimal retention. Instead, large sums are spent on retention bonuses. One Rand study[121] noted DOD's reenlistment bonus spending rose from $625 million in 2002 to $1.4 billion six years later.

Hardship pays are common, to include extra pays for locations overseas with below-standard quality of life. These range from $50 to $150 per month. However, these pays are not set dynamically (reflecting supply and demand). Though they rarely utilize it, the services have authority for a special pay that could radically reshape job-matching. It is known as the assignment incentive pay (AIP). According to the DOD, it is designed to encourage members to volunteer for difficult-to-fill jobs or assignments in less-desirable locations. The monthly statutory maximum payable is $3,000.

TABLE 6.5.
TVF Pay Table ($)

	TVF Baseline	+ Role	+ AIP	Max
O-8	10,000	0 to 10,000	0 to 3,000	23,000
O-7	8,500	0 to 9,000	0 to 3,000	20,500
O-6	6,500	0 to 8,000	0 to 3,000	17,500
O-5	5,000	0 to 7,500	0 to 3,000	15,500
O-4	4,200	0 to 7,000	0 to 3,000	14,200
O-3	3,500	0 to 6,000	0 to 3,000	12,500
O-2	3,000	0 to 5,000	0 to 3,000	11,000
O-1	2,300	0 to 5,000	0 to 3,000	10,300
E-9	4,200	0 to 7,500	0 to 3,000	14,700
E-8	3,500	0 to 7,000	0 to 3,000	13,500
E-7	3,000	0 to 6,000	0 to 3,000	12,000
E-6	2,300	0 to 5,000	0 to 3,000	10,300
E-5	1,800	0 to 5,000	0 to 3,000	9,800
E-4	1,600	0 to 5,000	0 to 3,000	9,600
E-3	1,500	0 to 5,000	0 to 3,000	9,500
E-2	1,400	0 to 5,000	0 to 3,000	9,400
E-1	1,200	0 to 5,000	0 to 3,000	9,200

Source: Author

The TVF would shift monthly compensation away from cumulative years of service and eschew tenure pays altogether. The alternative pay structure envisioned for the TVF is presented in table 6.5. Baseline pay is presented in a single column. The next column adds a monthly pay for the role: maintenance officer, cyber NCO, cryptographic/linguist, and so on for every occupational code across all the services. The role pay can also include different levels of responsibility, such as higher pay for command positions or other critical roles. The role pay could range from $0 to $10,000 per month. Included in these role pays is the level of responsibility, which can be much more nuanced than the tenure assumptions in the status quo. For example, there may be two USAF majors in the human intelligence (HUMINT) specialty, but one is a staff role at the Pentagon while the other is at a remote facility in the Middle East. The latter role would have a higher pay supplement, coded into the requirement. Moreover, the role pay would be publicly transparent, particularly visible on the online talent marketplace when the job is open.

The sum of these two pays—baseline and role—would serve as the base pay equivalent in current retirement formulas. TVF compensation should not be misconstrued as a way to stealthily slash Pentagon pension burdens. In addition, unit commanders would receive a 50 percent pay supplement during their months actively in command.

The role bonus would be composed of increments for skills and occupation (rather than skills alone). This authority would allow services to compensate the individuals who take on tougher jobs (including command) that involve higher career risk, longer hours, and greater stress. In principle, there is no reason to pay a senior O-3 in an easy job more than a junior O-3 in a demanding job. The same principle applies to E-5s, O-5s, E-6s, and so on.

The column "+AIP" stands for the assignment incentive pay, which already exists in current regulations, as mentioned earlier. The armed forces currently use bonus pays as flexible compensation for certain hazardous, remote, and otherwise challenging jobs. This flexibility is effective and should be expanded to enable decentralized job-matching in an evenhanded and objective manner to prohibit local favoritism. Jobs that remain unfilled after a given time should be paid more using a standardized incentive pay. As noted in chapter 3, wage flexibility is a core principle in labor markets.

The TVF recommendation is that the AIP should be used in conjunction with job-matching. The AIP should only be used objectively, not subjectively. That is, the AIP could not be granted at the discretion of a commander to award members of the unit. Rather, it would be applied according to an algorithm for unfilled requirements.

A posted requirement that receives no applicants for thirty days would get an automatic, publicly listed AIP for some percentage (say 5 percent) of base pay. The AIP would increase incrementally every fifteen days up to a dollar amount set by the service personnel command. Importantly, the AIP algorithm would reset to zero for each job posting. Patterns of unfilled jobs would show up in AIP data for roles throughout the workforce. Branch managers would be able to analyze the data and adjust the role supplements accordingly. The use of data—trending high and low labor volunteers over time—will allow compensation to be fine-tuned for roles throughout the military, generating untold cost savings and nimble force-shaping that are beyond reach today.

What if jobs remain unfilled even when the AIP has maxed out? After 180 days, qualifications should be relaxed to allow individuals one rank lower to apply.

There is one area where subjectivity should be allowed, and that is to lower or nix AIP for up to one-third of manpower requirements. The personnel center or local commander could use his judgment to designate non-AIP positions, giving a powerful signal regarding which ones are in fact high priority.

The consequence of shifting to TVF compensation would be a sea change in how force-shaping is done. Role pays could be adjusted nimbly to align hundreds of occupational labor supply-and-demand equilibria, from the half dozen Army helicopter repair specialties to the score of Air Force acquisition specialties. In the near term, monthly pay would not change much. In the long term, compensation would align with civilian peers in equilibrium, which will likely lower net costs to the military.

One last recommendation to change direct compensation is to reform the way travel for training and education programs is managed. A severe constraint on service flexibility is the Joint Federal Travel Regulation (JFTR), because it requires any program that lasts fewer than six months to be compensated with expensive "temporary duty" travel pay. Consequently, the services design most training programs to be longer than six months, which requires service members to incessantly move their permanent base locations. This regulation should be amended so that a greater variety of broadening programs can be offered. For example, active-duty troops could have the option to participate in congressional and business internships, or other brief training programs, without per diem reimbursement. To be clear, this would not mandate the end of temporary duty compensation, nor would it mandate any service branch to change its education and training programs; it would simply allow services more options to design and redesign training and education opportunities for members.

Specifically, the services should have the flexibility to waive TDY (temporary duty assignment) payments for optional training programs. Many graduate programs and fellowships are offered to senior officers as one-year assignments. Yet the same benefits—exposing officers to civilian colleges and private-sector firms—could be achieved with three- or four-month programs instead of nine or twelve months, which is the status quo. The roadblock is JFTR.

Transform the Military Pension

In January 2015, the Military Compensation and Retirement Modernization Commission issued its final report with fifteen recommendations for reform. The first recommendation was to lower the standard defined benefit pension by one-fifth and to add a 401(k)-style program. Unlike dozens of earlier reform commissions and boards, MCRMC achieved its goal. Congress was persuaded to develop legislation implementing the pension proposal, which it adopted later that year. Congress enacted a new Blended Retirement System—mandatory for new service members starting on January 1, 2018—which reduces the monthly DB payment from 50 percent of base pay to 40 percent. Technically, the base pay multiplier was reduced from 2.5 percent to 2 percent of base pay per year of service at the time of retirement. The blended plan also adds a supplemental defined contribution (DC) program.

MCRMC's proposals were a small step in the right direction, but policymakers should not consider its report as the last word or even a comprehensive analysis of alternatives. The commission's overall conclusion was that "several key features of the compensation system continue to meet the needs of the All-Volunteer Force."[122] I believe that is an incorrect assessment. The blended retirement is a step in the right direction, but it is a small step and is likely to be negligible in managing talent.

The standard military twenty-year cliff vesting creates a perverse incentive for both labor supply and labor demand, as the Gates commission noted decades ago. There are two main problems: the vesting date (at twenty) and the initial payment date (immediately upon retirement). The latter effectively pushes active-duty troops to immediately leave at the twenty-year point in order to begin drawing their pensions. And numerous studies have documented the reluctance of the military services to lay off senior officers in the years before the twenty-year point. This is the single greatest challenge for talent management, and it remains knotted.

A smarter approach would offer pension payments starting at the age of fifty-five or some length of time after the vesting date. Reservists get their pensions at age sixty irrespective of the age they retire. With the savings this extension will generate, monthly payments could be made larger. Another alternative would give entering service members the option to take higher blends of DC versus DB mixes, even allowing them to opt for full DC-only plans. MCRMC did not consider this, but it is a logical next step.

The latter half of this chapter highlights the inefficiency of the military DB structure using an analysis of the value of work. During the first year of service, annual base pay for a typical officer in any service is $35,212. By the twentieth year, base pay is $99,374. However, that last year of work will also qualify the individual for a lifetime stream of retirement income worth half of base pay, and the discounted value of that pension over a typical lifetime is $869,772. In other words, the twentieth year of service is worth a combined sum just under one million dollars.

What's Wrong with the Defined Benefit

The existing DB retirement is utilized by just one in six enlisted troops from all service branches. Roughly half of enlistees serve on active duty for ten years or less, and only 13 percent serve beyond twenty years of service. Those who serve for 20 years earn a generous retirement, but most do not stay in that long. In contrast, nearly half of all officers serve for twenty years or more, qualifying for the lifetime DB payments. That disparity alone is a hangover from the draft era, when enlistees were poorly paid, poorly valued labor. To be sure, young men are more naturally fit as warriors than older men, a reality that has not changed under the AVF and the force structure of the enlisted

corps and is particularly true for ground forces. Even today, the notion of a full career for an enlistee is encouraged only for the best who
are promoted to the middle and upper enlisted ranks as NCOs.

Another challenge is cost. In the coming years, the Pentagon is
likely to experience large pension pressures. Annual outlays for military pensions exceed $50 billion and will double during the next two
decades, while liabilities of the program are predicted to grow from
$1.3 trillion—roughly one-tenth the size of the US GDP—today to
$2.6 trillion in 2035, according to the Defense Business Board in 2011.
At best, the MCRMC reforms will shave one-fifth of the additional
pension burden, but it will take two decades for the next generation of
retirees to depart the service and the impacts of the reforms to be clear.

Long before a retirement benefit was the norm in the private sector, the military offered a simple deal: work for twenty years or longer
on active duty and in return receive half of your base salary after you
retire, for as long as you live. That's what came to be called a defined
benefit, and it was the norm in private-sector firms in the 1950s. However, the private sector shifted away from unsustainable DB plans
starting in the 1980s, partly because of cost and partly because they
lock in mediocre employees while nudging out entrepreneurial employees. More common today is a DC asset like the 401(k) that is
owned by individual workers.

The MCRMC final report acknowledged a widespread consensus:
"Services, as well as service members, would benefit from additional
flexibility in the compensation system." However, its blended proposal shifts the pension from being 100 percent DB to a blended plan
that has a DB/DC ratio of 80/20. And the underlying DB structure
remains the same, with no change at all to the timing of the vesting
cliff. There are four structural elements that could be adjusted: the
multiplier, vesting year, benefits initiation, and asset ownership. The
MCRMC alternative leaves three core structures unchanged, and
simply tweaks the multiplier (from 2.5 to 2 percent).

Importantly, MCRMC does not consider discontinuing the use of the DB pension. Instead, it asserts that the "current DB plan should be maintained because of its strong retention-pull effect on the Services' force profiles." This claim is not substantiated with references to research, scholarly or otherwise, merely a passing mention that it is the Pentagon's preference. Retention-pull of the current compensation structure is a core problem. To be sure, a DB pension is a magnet for *some* personnel, but which personnel? For undifferentiated labor, it doesn't matter, but for talent it is critical. Logic suggests that risk-averse officers are more attracted to a pension, and innovative officers are more attracted to a flexible, permeable force structure.

Consider the case of a highly valuable mid-career officer with expertise in cyber warfare. She is at the ten-year mark of her uniformed military service with a computer science degree, a passion for hacking, extensive military training, and valuable experience. According to the MCRMC's calculus, this officer's current DB package will be worth $712,000 if she stays an additional decade, which is about $70,000 per year. That amount is not a large incentive for a cyber-skilled officer. The point is not that such highly talented officers need more dollars to incentivize their service, because money is not the object. Career control and flexibility are the key incentives that talented individuals want, but will never have, with the existing DB superstructure. A larger point is that *efficient* force-shaping cannot be achieved in the modern era using twenty-year cliff vesting.

Rather than offer a set of alternatives—with analysis of trade-offs among them—the MCRMC offered one. To understand a fuller set of options and why they are necessary, it is helpful to consider some of the useful background published in the MCRMC report, as well as its *Interim Report* published in June 2014:

- The current US military pension is a defined benefit (DB) awarded to individuals who retire after twenty years of service.

Eighty-three percent of enlisted personnel and 51 percent of officers do not receive any pension because they leave before their twentieth year.

- The current DB formula sets one's monthly payment to the number of years of service (twenty-plus) times a multiplier of 2.5 percent of the average pay during one's highest three years of pay, plus a cost-of-living adjustment. As a rule of thumb, retirees receive 50 percent of their final paycheck (twenty years × 2.5 percent per year) for as long as they live.

- A larger portion of active-duty enlistees retires during the twenty-first year of service than during the previous seven to eight years combined. This retirement notch shows the strong incentive effect of the twenty-year DB.

- Pension vesting is twenty years for the US military pension but cannot exceed seven years (graduated) or five years (cliff) in the private sector. The reason is that vesting beyond seven years creates unjust monetary coercion by the employer on the employee. Long vesting also creates perverse incentives for both employer and employee to maintain a low-productivity employment relationship during latter years.

- The share of Americans with DB-only plans in the private sector fell from 62 percent to 7 percent between 1979 and 2009. One-third of American workers have blended plans that include DB and DC components. However, it is *not* clear if private DB plans are comparable to the military's in other ways: annuity payments, cost-of-living formulas, and so on.

- DB payments are received indefinitely for the life of the recipient. Personal savings pensions, known as 401(k) plans according to IRS tax law, and defined contribution (DC) plans more generally, carry more investment risk but are a genuine asset that can be inherited by spouses and other survivors when the recipient dies.

- Economic research shows that DB plans have a behavioral effect, leading to lock-in of some employees until vesting and a longer-term work disincentive after vesting.

The Value of Work

How much is the military's DB pension worth at the twenty-year mark? For the typical NCO, the discounted net present value is $201,282, according to MCRMC estimates.[123] The actual cost to the United States government would be significantly more, probably twice as much, but the perceived value for an individual who faces an uncertain future discounts the future dollars; hence this estimate is what the typical retiree would trade for if it were a lump sum. For a typical officer (lieutenant colonel/O-5) at the twenty-year point, MCRMC says the discounted net present value of the DB is closer to $712,000.

As the Gates Commission noted in its 1970 report, the twenty-year cliff pension distorts workplace incentives. A pension cliff of twenty years is illegal in a private-sector pension—three times longer than what is allowed—because it is so coercive as to be deemed abusive to the employee. Simple analysis reveals how strong and perverse the incentive is.

First, consider the value of work just before and after the retirement point. An NCO who enlisted at age eighteen can retire at the age of thirty-eight. More to the point, the newly retired NCO begins to draw a pension immediately. At the twenty-year mark, exactly, he qualifies for the DB pension, which is 50 percent of base monthly pay for the remainder of his life, whether he lives to thirty-nine or ninety-nine. Base monthly pay for an enlistee at the rank/pay grade E-8 at nineteen years of service is $4,878, which steps up to $5,009 at the twenty-year mark. At first blush, the value of work—that is, staying in uniform instead of retiring—is $5,009 per month. But two other factors must be considered.

Because the monthly pension payments would begin immediately upon retirement, his decision to *not* retire means forgone pension income. Working after twenty years on active duty is suddenly half as valuable as before: the net income benefit of working is half the gross amount in a full paycheck. The second, and more sizable, marginal impact of preretirement work is the *increasing* lifetime value of the DB pension for each additional month of work up to the twenty-year threshold. If an enlisted individual leaves the service after nineteen years and eleven months, he abandons a future stream of pension payments that could easily total a million dollars.[124] Because individuals discount the value of future payments, the perceived value of that pension income stream to the individual recipient is far less—I calculate it to be worth $533,000 to the typical veteran using a discount rate of 5 percent per year. It is worth even less to the enlistee on the first day of basic training due to an additional two decades of discounting, even if he plans on serving twenty years.

Individuals have what is called a present-bias for income for at least three reasons, and this is why individuals can be analyzed with a different discount rate than organizations or large groups. First, individuals face a risky future, especially a mortality risk. Suppose a veteran dies in an accident just three years after retiring: there is no future income stream. A second reason is that individuals are impatient, which is not a criticism and is the reason people use loans to purchase big-ticket items like automobiles, appliances, and houses. A third reason is uncertainty regarding inflation and other unforeseeable events. Large organizations can hedge against uncertainty in a way that individuals cannot.

Consider the perspective of the value of work for a master sergeant in the Air Force near the end of his twentieth year. His monthly base pay is $5,009, but the value of working during that last month rather than quitting is much more. The expected value of the last *month* of work in the US military before the twenty-year cliff is $538,284— base pay plus the entire expected value of the retirement package. Then the very next month's paycheck has a net marginal value of just over $2,504: his base pay minus the forgone retirement payment, which is half that amount. The incentive effects of these income differentials must be large.

Looking at the retirement cliff, the mid-career NCO calculates the average value of each month's work before and after retirement with the pension as the primary decision factor. With ten months until retirement, each month is worth about one-tenth of its retirement value, or $50,000. With five months to go, each month is worth twice that. The average monthly value escalates until the twenty-year vesting threshold is reached.

At the threshold of twenty years in uniform, for everyone on active duty, the economic value of work collapses. This probably explains why over half of the retirees from every branch of service choose that moment to retire.

In my analysis here, I use some strong simplifying assumptions to model the cost of the DB pension versus the value to the retiree. I assume an average lifespan of eighty years. I assume inflation is zero and cost-of-living adjustments are zero. I assume that the government earns zero interest when it sets aside dollars to pay future pensions. Most importantly, I assume that individuals have a discount rate of 5 percent more than the government's because the government faces that much less risk and uncertainty. Most scholars treat individual discount rates as equivalent to institutional discount rates, so my assumption of a discount gap is driving a large difference in the perceived value that veterans see in their retirement pay and how the Pentagon sees it.

This means that future income streams are discounted at a rate of 5 percent each year, with projected income streams in future decades worth less than the stream in the present. For example, Charlie would value 95 cents today equally with a promise to pay Charlie 100 cents a year from today. By using a different discount, I am suggesting that the cost of the government offering an annuity is much more than the benefit an individual derives from getting one.

Monthly retirement pay for the current retiring officer at twenty years is $49,874 (this is half of the average of the highest three years of pay). Assuming a life expectancy of eighty-four years and a retirement age of forty-one years, I calculated the discounted value of that income stream at the moment of retirement to be $925,163. The cost to the government is, I assume, not discounted and therefore stands at $2.14 million.

To understand the powerful incentive of the military's lifetime defined benefit, I calculated the value of a year's work for the typical officer over twenty-four years of a career. Figure 6.3 shows how this value grows over time, in which the value of serving each year is that year's base pay plus the total value of a DB pension divided by the years remaining until vesting. For example, two years before retirement, the total value of working that year is $520,000 ($100,668 base pay plus half of the discounted lifetime value of the pension, which at

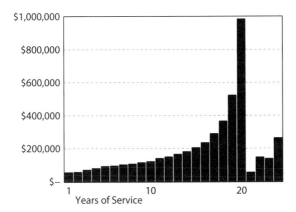

FIGURE 6.3. Officer's Value of Work by YOS (pay plus pension)
Source: Author's calculations

that moment is $839,150). One year before retirement, a year's work is worth just under $1 million to the typical American officer.

There is a collapse of work value at year twenty-one. Monthly base pay is suddenly half as valuable as before—the *net* income of working is half the gross amount in a full paycheck. The incentive effects almost certainly explain the twenty-year retirement bubble. It is also notable how quickly the value of each additional year of work rises, assuming promotion to O-6 at year twenty-two, and this is because pay rises dramatically (and consequently so does the pension value) with years of tenure between twenty and twenty-eight years of service, regardless of role, responsibility, or rank.

This raises the question of whether the new blended system will smooth out the twenty-year *value of work*, and the answer is no (see figures 6.4 and 6.5, based on calculations I made for this book). The new blended retirement program does not change the shape of the work value ramp at all, only its peak, which will be 20 percent lower because of the smaller DB and only slightly higher because of the new asset component (adding 5 percent of matched base pay).

In the final analysis, there are two critiques of the current military pension. One is that the perceived value of the pension to individual veterans is far lower than its cost to the government. I made a number of assumptions to make that critique, but the one that really matters is the gap in discount rates. Even if you do not buy that argument, the second critique is that the twenty-year cliff severely distorts work incentives. There are no assumptions needed, because the shape of the value of work over time is the same, regardless of discount rates and cost-of-living adjustments.

TVF Alternatives

A well-designed pension would aim to optimize management of talent by flattening the annual value of work, primarily to avoid expo-

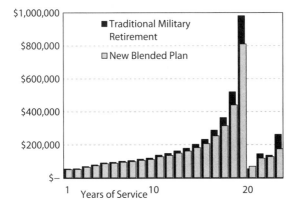

FIGURE 6.4. Officer's Value of Work by YOS (pay plus pension), Comparing Pension Plans
Source: Author's calculations

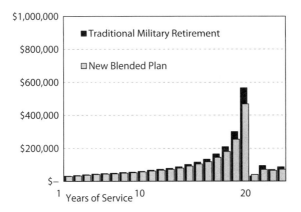

FIGURE 6.5. NCO's Value of Work by YOS (pay plus pension), Comparing Pension Plans
Source: Author's calculations

nential rising threshold spikes in pensions or other forms of delayed compensation. This section introduces two alternative designs for a military pension that would better optimize permeability while lowering overall costs to the military.

The first alternative retains the DB structure, but shifts vesting to year ten and initial payment out to age fifty-five. The second alternative ends the use of DB payments in lieu of a full DC savings asset.

Legislators and military leaders could avoid concerns about overly radical reforms by retaining the status quo as an option. Entering service members should be given a menu of the existing blended DB or one of the two TVF alternatives. In addition, service chiefs should be empowered to select one of the alternatives for their new accessions. There is no reason that the Navy should be required to offer the same retirement package as the Air Force. Each service has unique talent needs across multiple dimensions (age distribution, physical fitness, rank structure, and occupational skills, to name a few) that call for tailored compensation structures rather than a one-size-fits-all plan. Table 6.6 summarizes the key design features of each plan, as compared to the two plans in effect now.

The "TVF DB" is a modernized defined benefit, with a far less coercive vesting cliff of ten years, whereas the payout of benefits begins over a decade later at the age of fifty-five (whether the member is retired at that point or not). It pushes out the date pension payments begin for two reasons: to end coercive incentives and to save money. However, an earlier vesting YOS will incur major new costs for the younger service members who will qualify for it. In order to keep a lid on those costs, the services should change incentives only gradually. The multiplier need not be the same for every year served. TVF DB is modeled with an initial multiplier of 1 percent for years one to ten, 2 percent for years eleven to fifteen, and 3 percent for every year thereafter. The discounted value of the TVF DB pension for a twenty-year retiree is three-quarters of the status quo, even though the full cost to the government is slightly more than half of the status quo.

I also modeled the value and cost of a completely new kind of military pension along the lines called for by the Defense Business Board. This "TVF Save" savings plan is a DC-only plan. It would discontinue the DB plan completely for new officers and enlistees. Any full DC plan allows for greater flexibility in managing the force, allowing

LE 6.6.
tary Officer Retirement and TVF Alternatives (at 20 YOS)

	DB (status quo)	Blended (status quo)	TVF DB	TVF Save
tiplier	2.5%	2.0%	1.0 to 3.0%	—
ting at (year of svc)	20	20	10	—
benefits begin	41	41	55	—
ual pension pay, 42–54	$49,874	$39,899	$-	—
ual pension pay, 55+	$49,874	$39,899	$34,912	—
value to retiree	$925,163	$856,634	$647,614	$905,819
cost to US Military	$2,144,582	$1,786,620	$1,047,354	$817,421
ings asset at 10 YOS	$-	$30,560	$-	$154,047
et (personal property)	No	Some	No	Yes
k incentive	No	No	Yes	Yes
nt Management	No	No	Some	Yes

ce: Author's calculations

individuals to take off-ramps and on-ramps between active duty, reserve status, and the private sector.

The plan I modeled begins contributions during the third year and scales up employer matching from 1 percent to 20 percent during the first five years. A generous direct employer contribution that equals 25 percent of base pay per year is added to the personal savings account beginning in year six, plus a match of funds saved by the individual up to a maximum of 25 percent of base pay. If members save the maximum amount, the Pentagon would contribute 50 percent of base pay to the savings account monthly—far in excess of the private-sector norm. Yet it would cost the military far less than the status quo. Under this plan, service members create an asset that is valued by the twenty-year retiree at the *same* level as the status quo DB pension, meanwhile saving the government $1.3 million per retiree.

The TVF Save plan would incur additional costs as well, due to the number of service members who leave between six and twenty years and will receive a generous savings asset. It is impossible to model the net savings or cost military-wide, but overall costs of this plan must be considered in balance with increased permeability and therefore overall compensation that will fall to a lower equilibrium due to increased labor supply.

Value of work under the two TVF plans is radically different than the two status quo plans. Members will be likely to leave the service more frequently in the absence of twenty-year vesting. However, these new pension structures also enable a wide-scale continuum of service careers with frequent reentry of reservists and other veterans to active-duty status. The military stands to enhance its talent pool with continuum careers, and it will also generate a much thicker labor supply than it currently has—one that it can be selective in accessing.

Additional Background

The January 2015 final report from the MCRMC is not the first major government assessment to recommend changing the US military's compensation structure. A few years after the AVF was in place, on March 13, 1978, the US Government Accountability Office (GAO) was calling for an end to the twenty-year cliff in a clearly titled report, *Retirement Security: The 20-Year Military Retirement System Needs Reform:*

> Twenty-year retirement, in conjunction with present personnel management policies, is an inefficient means of attracting new members, causes the services to retain more members than are needed up to the 20-year point, provides too strong an incentive for experienced personnel to leave after serving 20 years, and makes it impossible for the vast majority of members to serve full careers.

This echoed the findings of the *President's Commission on Military Compensation* (PCMC) in 1978, which, in the words of a recent Rand summary, "criticized the military retirement system on three counts: that it was inequitable because there were no retirement benefits for those not reaching 20 YOS; that it *inhibited flexible force management because managers were reluctant to separate low performers as they neared 20 YOS* (emphasis added); and that it was ineffective because it had little effect on recruiting and early retention, but an extremely strong effect on retention after 10 or 12 YOS."

On November 15, 1996, the GAO issued a report titled *Military Retirement: Possible Changes Merit Further Evaluation*. From the summary:

The military retirement system strongly influences the broad shape of the force, since it provides an increasing incentive for service members to stay in the military as they approach 20 years of service and encourages them to leave thereafter, helping DOD to retain midcareer personnel and yielding a relatively young force; (6) however, the system can also impede effective force management because military personnel are not entitled to any retirement benefits unless they have served 20 years, and the *services have been reluctant to involuntarily separate personnel* with less than 20 years of service beyond a certain point due to the financial consequences for service members and the impact on morale; (7) some analysts, including several of GAO's roundtable participants, believe the military retirement system is an obstacle to achieving the right force size and composition. (emphasis added)

In 2006, the Defense Advisory Committee on Military Compensation (DACMC) issued its report, which took aim at the status quo pension system as inefficient, inflexible, and inequitable. It proposed a hybrid plan, what many scholars call a "blended" plan, with DB and DC components. Its DB component would retain the 2.5 multiplier, vest earlier (year ten), but pay out at age sixty. Once again, the underlying

motivation cited by DACMC was the need for flexibility in force management. In 2011, the Defense Business Board issued a report that offered the latest harsh critique and proposal for pension reform.

Indeed, before the AVF was implemented in 1973, the Gates Commission recommended a series of personnel reforms that it considered
necessary. Its proposal called for an introduction of "vesting" over time
in contrast to the binary all-or-nothing cliff at twenty years. The incentive effect of the pension is carefully discussed in the following passage,
something lacking altogether in the hundreds of pages of the MCRMC's
final report. Here is the Gates Commission:

> Because military retirement benefits are budgeted each year out of current
> funds, servicemen have never acquired vested retirement rights except those
> which arise after one has served long enough (19½ years) to be eligible to retire.
> This policy has a number of undesirable effects. . . .
>
> Retirement benefits have the additional disadvantage of being worth too
> much to the individual who is beyond his tenth or eleventh year of service. He
> cannot afford to separate from the service because of the benefits for which he
> will qualify after another 9 or 10 years. Because of this potential loss, the mili-
> tary rarely discharges individuals who have served more than say, 10 years.
>
> Also, the substantial retirement income available after 20 years of service
> (when some enlisted men are only 37 years old) induces many individuals to re-
> tire as soon as they are eligible. The combination of retirement income and civil-
> ian earnings is very attractive. Men who retire early are often those with superior
> civilian earning opportunities and they are precisely the individuals the services
> would like to retain longer. The Department of Defense group organized to study
> compensation recognized the shortcomings of the present retirement system.

Conclusion

The Total Volunteer Force would reform military compensation by
transforming base pay and retirement. Reforms to the military pen-

sion would not affect the retirement payments and promises to veterans or troops currently on active duty. Instead, their impact would be limited to future service members with the simple exception being that service chiefs should have the authority to allow their service members the option to transition to one of the new options. As for transforming base pay, this chapter has shown that the long-standing pay tables are outdated. The roots of the tables date back to the beginning of the republic, but US military pay tables in the mid-nineteenth century had more occupational flexibility and less tenure rigidity than in 2016. A move to role-based pay is long overdue, and the ability to use assignment-pay incentives already exists in law.

This book makes the case that the US military offers many lessons, positive and negative, about organizational design. The final three chapters focus on three areas of talent management where the armed forces are self-graded as deficient: job-matching, performance evaluations, and compensation. Yet it would be an error to look to the private sector for best practices. Indeed, it was exactly this kind of hubris in the 1960s when Robert McNamara centralized HR policies at the Pentagon that created the problems that persist today. Unfortunately, the centralized rules and regulations have become cultural norms, and reforms to them are seen as taboos violating an ancient trust. Fortunately, the organizational assessment presented in chapter 1 revealed that the US armed forces are world-class in terms of most leadership metrics, particularly the values and sense of purpose that are unique as well as sacred.

Other firms would do well to study how the US military builds those bonds of trust, self-sacrifice, and voluntary service despite the red tape. It is the spirit of volunteerism inherent in this new generation of senior officers that is already breaking through the bureaucratic concrete. We can look to the world-class performance evaluations pioneered by the Marines or the peer ratings done by the Army Rangers to find hope. After a decade at war, American troops are coming home with little tolerance for regulatory barriers to excellence.

Leader/Talent Categories and Elements

Leadership Categories and Elements

1. Independence

- Individual leaders are empowered to use their own judgment.
- Leaders are willing and encouraged to take risks.
- Creativity is rewarded more than conformity.

2. Development

- Young leaders are given serious responsibilities.
- All workers are given the opportunity to truly excel.
- There are many great leaders in the organization.
- Mentors and mentoring are common.
- People get honest, informal feedback from bosses.

3. Purpose

- People are passionate about their work.
- This organization has a strong sense of purpose.

- Leaders are very motivational.
- Top leaders communicate a clear and compelling vision.

4. Values

- Teamwork is strong throughout the entire organization.
- There is a high degree of trust among bosses, peers, and subordinates.
- Personal integrity is valued, and low integrity is not tolerated.
- People sacrifice for each other within and across departments.

5. Adaptability

- The focus here is on mission success, regardless of barriers.
- Leaders are dynamic: able to change direction when the mission changes.
- Bureaucratic rules are streamlined and do not get in the way.

Management Categories and Elements

1. Training

- Major occupational training is valuable.
- On-the-job training is valuable.
- The organization avoids excessive management briefings and meetings.
- The succession process, when a new manager fills an existing job, tends to be seamless.

2. Job-Matching

- Local supervisors have primary hiring authority, not the central bureaucracy.
- Poorly performing employees can be easily removed, relocated, or fired.

- Job requirements and personal skills are usually well matched.
- Jobs are flexibly redefined whenever the operating environment changes.

3. Compensation

- Pay is closely aligned with performance.
- Bonuses are used effectively to reward good work.
- Fringe benefits are efficiently set, but not wasted or lavish.
- Retirement and retention programs help keep top talent and enhance the long-term success of the organization.

4. Evaluations

- Performance evaluations provide valuable information for employee feedback.
- Performance evaluations provide valuable information for job assignments and promotions.
- Assessments of supervisors by their peers and subordinates are considered by senior executives.
- Performance evaluations make useful distinctions between top, middle, and weak performers.
- A respected program of commendations, medals, and/or awards exists to recognize top performers.

5. Promotions

- Promotions are based on merit, not on seniority or favoritism.
- Individuals are free to specialize rather than seek managerial promotions.
- Abusive bosses are not tolerated and are removed.
- Great talent is promoted quickly.

Bibliography

Aguinis, Herman, and Ernest O'Boyle Jr. "The Best and The Rest: Revisiting the Norm of Normality of Individual Performances." *Personnel Psychology* 65, no. 1 (Spring 2012): 79–119.

Asch, Beth J., James R. Hosek, and John T. Warner. "New Economics of Defense Manpower in the Post–Cold War Era." In *The Handbook of Defense Economics,* vol. 2, edited by Todd Sandler and Keith Hartley, 1075–1138. Oxford, UK: Elsevier, 2007.

Barno, David. "Military Brain Drain." *Foreign Policy*, February 13, 2013, http://foreignpolicy.com/2013/02/13/military-brain-drain/.

Barno, David, and Nora Bensahel. "Can the U.S. Military Halt Its Brain Drain?" *The Atlantic*, November 5, 2015, www.theatlantic.com/politics/archive/2015/11/us-military-tries-halt-brain-drain/413965/.

Barno, David, Nora Bensahel, Katherine Kidder, and Kelley Sayler. *Building Better Generals.* Center for a New American Security, October 28, 2013, www.cnas.org/publications/reports/building-better-generals.

Baron, James N., and David M. Kreps. *Strategic Human Resources: Frameworks for General Managers.* Hoboken, NJ: John F. Wiley & Sons, 1999.

Bass, Bernard M., and Bruce J. Avolio. "Transformational Leadership and Organizational Culture." *Public Administration Quarterly* 17, no. 1 (Spring 1993): 112–21.

Bennis, Warren. *On Becoming a Leader,* 4th ed. Philadelphia: Basic Books, 2009.

Bloom, Nicholas, Renata Lemos, Raffaella Sadun, Daniela Scur, and John Van Reenen. "The New Empirical Economics of Management." *Journal of the European Economic Association* 12, no. 4 (August 2014): 835–76.

Brett, Joan F., and Leanne E. Atwater. "360 Degree Feedback: Accuracy, Reactions, and Perceptions of Usefulness." *Journal of Applied Psychology* 86, no. 5 (2001): 930–42.

Carter, Ashton. "Remarks by Secretary Carter on the Force of the Future." Speech at Abington Senior High School, Abington, Pennsylvania, March 30, 2015, www.defense.gov/News/Speeches/Speech-View/Article/606658.

Colarusso, Michael J., and David S. Lyle. *Senior Officer Talent Management: Fostering Institutional Adaptability*. Carlisle, PA: US Army War College Press, 2014.

Congressional Budget Office. "The All-Volunteer Military: Issues and Performance." Publication 2960, July 19, 2007, www.cbo.gov/sites/default/files/110th-congress -2007-2008/reports/07-19-militaryvol_0.pdf.

Coumbe, Arthur T. "Army Officer Development: Historical Context." Strategic Studies Institute, US Army War College, 2010.

Defense Business Board. "Modernizing the Military Retirement System." Report FY11-05, Washington, DC, October 2011, http://dbb.defense.gov/Portals/35 /Documents/Reports/2011/FY11-5_Modernizing_The_Military_Retirement _System_2011-7.pdf.

Defense Science Board. "Enhancing Adaptability of U.S. Military Forces," January 2011, www.acq.osd.mil/dsb/reports/EnhancingAdaptabilityOfUSMilitary ForcesB.pdf.

Dempsey, Martin E. "Building Critical Thinkers: Leader Development Must Be the Army's Top Priority." *Armed Forces Journal*, February 1, 2011, http:// armedforcesjournal.com/building-critical-thinkers/.

Department of the Army. "Commissioned Officer Professional Development and Career Management." Pamphlet 600–603, February 1, 2010.

Dessler, Gary. *Human Resource Management*, 12th ed. Upper Saddle River, NJ: Prentice Hall, 2011.

Drucker, Peter F. *Management Challenges for the 21st Century*. New York: Harper-Collins, 2001.

Fowler, Floyd J., Jr. *Survey Research Methods*, 4th ed. Thousand Oaks, CA: Sage, 2008.

Gates, Robert M. "Secretary of Defense Speech." US Military Academy, West Point, NY, February 25, 2011, http://archive.defense.gov/Speeches/Speech .aspx?SpeechID=1539.

Government Accountability Office. "What Defense Says about Issues in Defense Manpower Commission Report: A Summary." May 3, 1977.

Government Accountability Office. "The 20-Year Military Retirement System Needs Reform." FPCD-77-81, March 13, 1978, www.gao.gov/products/FPCD -77-81.

Government Accountability Office. "Military Retirement: Possible Changes Merit Further Evaluation." NSIAD-97-17, November 15, 1996, www.gao.gov /products/NSIAD-97-17.

Government Accountability Office. "Military Personnel: Strategic Plan Needed to Address Army's Emerging Officer Accession and Retention Challenges." GAO-07–224, January 19, 2007, www.gao.gov/new.items/d07224.pdf.

Halter, Scott M. "What Is an Army but the Soldiers? A Critical Assessment of the Army's Human Capital Management System." *Military Review* 92, no. 1 (January–February 2012): 16–23.

Hanushek, Eric A. "The Volunteer Military and the Rest of the Iceberg." *Policy Sciences* 8, no. 3 (September 1977): 343–61.

Holbrook, Allyson, Jon Krosnick, and Alison Pfent. "The Causes and Consequences of Response Rates in Surveys by the News Media and Government Contractor Survey Research Firms." In *Advances in Telephone Survey Methodology*, edited by James M. Lepkowski, Clyde Tucker, J. Michael Brick, Edith D. De Leeuw, Lilli Japec, Paul J. Lavrakas, Michael W. Link, and Roberta L. Sangster. Hoboken, NJ: Wiley, 2007.

Hosek, James, and Beth Asch. "Air Force Compensation: Considering Some Options for Change." Rand Corporation, 2002, www.rand.org/pubs/monograph _reports/MR1566-1.html.

Kane, Tim. *Bleeding Talent: How the US Military Mismanages Great Leaders and Why It's Time for a Revolution*. London: Palgrave Macmillan, 2012.

Kane, Tim. "How to Lose Great Leaders? Ask the Army." *Washington Post*, February 5, 2013, www.washingtonpost.com/national/on-leadership/how-to-lose-great-lead ers-ask-the-Army/2013/02/05/725f177e-6fae-11e2-ac36-3d8d9dcaa2e2_story .html.

Kane, Tim. "The Leader/Talent Matrix: An Empirical Perspective on Organizational Culture." Economics Working Paper No. 15106. Hoover Institution, June 2, 2015.

Kidder, Katherine, and Phillip Carter, *Military Compensation and Retirement Modernization: A Primer*. Center for a New American Security, January 2015, www.cnas.org/publications/reports/military-compensation-and-retirement -modernization-a-primer.

Lazear, Edward P., and Michael Gibbs. *Personnel Economics in Practice*, 2nd ed. Hoboken, NJ: John F. Wiley & Sons, 2009.

McCormick, David. *The Downsized Warrior: America's Army in Transition*. New York: New York University Press, 1998.

Michaels, Ed, Helen Handfield-Jones, and Beth Axelrod. *The War for Talent*. Boston: Harvard Business Review Press, 2001.

Moran, William. "A Conversation with Vice Admiral William Moran." By K. T. McFarland. Council on Foreign Relations, December 9, 2014, www.cfr.org /defense-and-security/conversation-vice-admiral-william-moran/p34004.

Mulligan, Casey B., and Andrei Shleifer. "Conscription as Regulation." *American Law and Economics Review* 7, no. 1 (2005): 85–111.

O'Reilly, Charles, and Jeffrey Pfeffer. *Hidden Value: How Great Companies Achieve Extraordinary Results with Ordinary People.* Cambridge, MA: Harvard Business School Press, 2000.

O'Reilly, Charles A., III, David F. Caldwell, Jennifer A. Chatman, and Bernadette Doerr. "The Promise and Problems of Organizational Culture: CEO Personality, Culture, and Firm Performance." *Group & Organization Management* 39, no. 6 (2014): 595–625.

Rostker, Bernard. *I Want You! The Evolution of the All-Volunteer Force.* Santa Monica, CA: Rand Corporation, 2006.

Rostker, Bernard D. *Reforming the American Military Officer Personnel System.* Testimony before United States Senate Committee on Armed Services, December 2, 2015.

Rostker, Bernard D., Harry Thie, James Lacy, Jennifer Kawata, and Susanna Purnell. *The Defense Officer Personnel Management Act of 1980: A Retrospective Assessment.* Santa Monica, CA: Rand Corporation, 1993, www.rand.org/pubs/reports/R4246 .html.

Ulmer, Walter F., Jr. "Military Leadership into the 21st Century: Another 'Bridge Too Far'?" *Parameters* 38 (Spring 1998): 135–55.

Wardynski, Casey, David S. Lyle, and Michael J. Colarusso. "Towards a U.S. Army Officer Corps Strategy for Success: Retaining Talent." Strategic Studies Institute, US Army War College, January 2010, www.strategicstudiesinstitute .army.mil/pubs/display.cfm?pubID=965.

Endnotes

1. Robert M. Gates, "Secretary of Defense Speech," US Military Academy, West Point, NY, February 25, 2011, http://archive.defense.gov/Speeches/Speech .aspx?SpeechID=1539.

2. Ash Carter, "Remarks by Secretary Carter on the Force of the Future," speech at Abington Senior High School, Abington, PA, March 30, 2015, www. defense.gov/News/Speeches/Speech-View/Article/606658/remarks-by-secretary-carter-on-the-force-of-the-future.

3. Andrew Tilghman, "The Army's Other Crisis: Why the Best and Brightest Young Officers Are Leaving," *Washington Monthly,* December 2007.

4. Bernard Rostker and K. C. Yeh, *I Want You! The Evolution of the All-Volunteer Force* (Santa Monica, CA: Rand Corporation, 2006), 66.

5. Gates Commission Report, letter of transmittal from Thomas S. Gates to the president, February 20, 1970.

6. See Carter, "Remarks by Secretary Carter on the Force of the Future."

7. See US Department of Labor, Bureau of Labor Statistics, "Employment Situation of Veterans—2015," news release, March 22, 2016, www.bls.gov/news .release/pdf/vet.pdf.

8. Ibid. Also see table 10 of the BLS "Employment Situation of Veterans" report, www.bls.gov/news.release/vet.t10.htm.

9. Department of Veterans Affairs, *2015 Veteran Economic Opportunity Report,* www.benefits.va.gov/benefits/docs/VeteranEconomicOpportunityReport2 015.PDF.

10. See Jesse Sloman, "The U.S. Military's Next Big Reform Challenge Is Here," *National Interest*, January 5, 2015, http://nationalinterest.org/blog/the-buzz/the-us-militarys-next-big-reform-challenge-here-11970.

11. Peter F. Drucker, *Management Challenges for the 21st Century* (New York: HarperCollins, 2001); Warren Bennis, *On Becoming a Leader*, 4th ed. (Philadelphia: Basic Books, 2009).

12. Tim Kane, *Bleeding Talent: How the US Military Mismanages Great Leaders and Why It's Time for a Revolution* (London: Palgrave Macmillan, 2012); Casey Wardynski, David S. Lyle, and Michael J. Colarusso, "Towards a U.S. Army Officer Corps Strategy for Success: Retaining Talent," Strategic Studies Institute, US Army War College, January 2010, www.strategicstudiesinstitute.army.mil/pubs/display.cfm?pubID=965; Arthur T. Coumbe, "Army Officer Development: Historical Context," Strategic Studies Institute, US Army War College, 2010; David Barno, Nora Bensahel, Katherine Kidder, and Kelley Sayler, "Building Better Generals," Center for a New American Security, October 2013, https://s3.amazonaws.com/files.cnas.org/documents/CNAS_BuildingBetterGenerals.pdf.

13. William Moran, "A Conversation with Vice Admiral William Moran," by K. T. McFarland, Council on Foreign Relations, December 9, 2014, www.cfr.org/defense-and-security/conversation-vice-admiral-william-moran/p34004.

14. Nicholas Bloom, Renata Lemos, Raffaella Sadun, Daniela Scur, and John Van Reenen, "The New Empirical Economics of Management," *Journal of the European Economic Association* 12, no. 4 (August 2014): 835–76.

15. See Tim Kane, "The Leader/Talent Matrix: An Empirical Perspective on Organizational Culture," Economics Working Paper 15106, Hoover Institution, June 2, 2015, www.hoover.org/sites/default/files/leadertalent.pdf.

16. Kane, *Bleeding Talent*.

17. This is the score for each of the individual services, although those are not reported here nor tested for significance.

18. James Fallows, "The Tragedy of the American Military," *The Atlantic*, January/February 2015.

19. Ash Carter, "Remarks on 'Building the Force for the Future,'" speech at George Washington University Elliott School of International Affairs, Washington, DC, November 18, 2015, www.defense.gov/News/Speeches/Speech-View/Article/630415/remarks-on-building-the-first-link-to-the-force-of-the-future-george-washington.

20. Thomas Ricks, "There's a Reason America's Vets Can't Find Work but Not What Ben Bernanke Thinks," *Foreign Policy*, October 5, 2015, http://foreignpolicy.com/2015/10/05/theres-a-reason-americas-vets-cant-find-work-but

-not-what-ben-bernanke-thinks/; David S. Loughran, "Why Is Veteran Unemployment So High?" Rand Corporation, 2014, www.rand.org/pubs/research_reports/RR284.html.

21. Department of Veterans Affairs, *2015 Veteran Economic Opportunity Report*.

22. US Department of Labor, Bureau of Labor Statistics, "Business Employment Dynamics Summary," news release, November 9, 2016, www.bls.gov/news.release/cewbd.nr0.htm.

23. US Department of Labor, Bureau of Labor Statistics, "Number of Jobs Held, Labor Market Activity, and Earnings Growth among the Youngest Baby Boomers: Results from a Longitudinal Survey," news release, March 31, 2015, www.bls.gov/news.release/pdf/nlsoy.pdf.

24. Department of Defense: Military Programs, *Office of Management and Budget* (February 2015): 233–34, www.whitehouse.gov/sites/default/files/omb/budget/fy2015/assets/mil.pdf; Office of the Undersecretary of Defense, "Fiscal Year 2016 Budget Request," February 2015. Note president's budget request and civilian full-time equivalents, http://comptroller.defense.gov/Portals/45/Documents/defbudget/fy2016/FY2016_Budget_Request.pdf.

25. Department of Defense: Military Programs (February 2015): 233.

26. "Population Representation in the Military Services: Fiscal Year 2013 Summary Report," CNA Corporation, for the Office of Undersecretary of Defense, Personnel and Readiness, 2013, 3.

27. "U.S. Army Weight Requirements," Military.com, www.military.com/join-armed-forces/army-weight-rules.html.

28. "Population Representation in the Military Services," 2–3.

29. "2014 Demographics: Profile of the Military Community," US Department of Defense, 2014, 21.

30. "2014 Demographics: Profile of the Military Community," 48.

31. Shanea Watkins and James Sherk, "Who Serves in the U.S. Military? The Demographics of Enlisted Troops and Officers," Center for Data Analysis Report no. 08-05 on National Security and Defense, Heritage Foundation, August 21, 2008. This is an updated study from the original 2005 study, which found 14.6 percent of recruited enlistees from low-income families and 22.2 percent from high-income families.

32. Ellen Turner, "A Profile of the U.S. Army: A Reference Handbook 2014/2015," Institute of Land Warfare, 2014, 25, www.ausa.org/publications/ilw/digitalpublications/documents/profile2014/index.html.

33. "Remarks by the Honorable Ray Mabus, Secretary of the Navy," Advance Bioeconomy Leadership Conference, March 11, 2015, 3, www.Navy.mil/Navydata/people/secnav/Mabus/Speech/ABLC2015.pdf.

34. US Department of Defense, "Quadrennial Defense Review 2014," 40.

35. Quadrennial Defense Reviews for 2001, 2010, and 2014; President's Budget Requests FY 1995–2017, for example the FY 2016 at www.whitehouse.gov/sites/default/files/omb/budget/fy2016/assets/budget.pdf; James C. Ruehrmund and Christopher J. Bowie, "Arsenal of Air Power: USAF Inventory 1950–2009," a Mitchell Institute study, November 2010, 14–26; Naval History and Heritage Command, US Ship Force Levels 1886–Present, www.history.navy.mil/research/histories/ship-histories/us-ship-force-levels.html#2000; Dakota L. Wood, ed., "2015 Index of U.S. Military Strength," Heritage Foundation, 304–5; Dakota L. Wood, ed., "2016 Index of US Military Strength," Heritage Foundation, 308–9; United States Air Force Budget Fiscal Year 2016 Overview, 37, www.saffm.hq.af.mil/Portals/84/documents/FY16/AFD-150421-011.pdf?ver=2016-08-24-100152-050.

36. Cindy Williams, "Transforming the Rewards of Military Service," *MIT Security Studies Program Occasional Paper,* September 2005: 24, www.comw.org/od/fulltext/0509williams.pdf.

37. Susan Landau, "What We Must Do about Cyber," *Lawfare* (blog), March 10, 2015, www.lawfareblog.com/what-we-must-do-about-cyber.

38. United States Office of Management and Budget Request FY 2017 (February 2016), 5-5, www.whitehouse.gov/sites/default/files/omb/budget/fy2017/assets/budget.pdf. Note: $6.7 billion for FY 2017 is an increase of $0.9 billion from FY 2016. According to FY 2016 President's Budget Request (February 2015): 5-5, USCYBERCOM is scheduled for completion in 2018.

39. Steven Aftergood, "Pentagon's Cyber Mission Force Takes Shape," *Secrecy News* (blog), Federation of American Scientists, September 10, 2015, https://fas.org/blogs/secrecy/2015/09/DOD-cmf/.

40. Aliya Sternstein, "US Cyber Command Has Just Half the Staff It Needs," *Defense One,* February 2015, www.defenseone.com/threats/2015/02/us-cyber-command-has-just-half-staff-it-needs/104847/; see also Mark Pomerleau, "What Will the Cyber Mission Force Look Like?" *Defense Systems,* October 13, 2015, https://defensesystems.com/articles/2015/10/13/us-cyber-command-cyber-mission-force.aspx.

41. Aftergood, "Pentagon's Cyber Mission Force Takes Shape."

42. United States Office of Management and Budget Request FY2017: 8–24.

43. Bill Canis, "Unmanned Aircraft Systems (UAS): Commercial Outlook for a New Industry," Congressional Research Service, R44192, September 9, 2015, 6, www.fas.org/sgp/crs/misc/R44192.pdf. Specifically, the USAF owns 34 percent, the Navy owns 36 percent, and the Marine Corps has 2.4 percent.

44. US Air Force, Fact Sheets on MQ-1B Predator, MQ-9 Reaper, and RQ-4 Global Hawk, current as of 2015, www.af.mil/AboutUs/FactSheets/tabid /131/Indextitle/R/Default.aspx.

45. United States Office of Management and Budget Request FY 2016: 8–16. The new plan would require the Air Force to fly sixty CAPs a day; Army up to sixteen; Special Forces up to four; and contractors up to ten.

46. United States Office of Management and Budget Request FY 2016: 8–16. See also Mark Pomerleau, "Air Force Addresses Drone Pilot Shortage," *Defense Systems*, May 20, 2015, https://defensesystems.com/articles/2015/05/20/air -force-addresses-drone-pilot-shortage.aspx.

47. US Department of Defense, "Department of Defense (DoD) Releases Fiscal Year 2017 President's Budget Proposal," news release, February 2016, www .defense.gov/News/News-Releases/News-Release-View/Article/652687 /department-of-defense-DOD-releases-fiscal-year-2017-presidents-budget -proposal.

48. Bradley T. Hoagland, "Manning the Next Unmanned Air Force: Devel- oping RPA Pilots of the Future," Brookings Institution, August 2013, www .brookings.edu/research/papers/2013/08/06-air-force-drone-pilot-develop ment-hoagland. Of 150 positions, 123 were filled.

49. Ibid., 2.

50. Ibid., iv.

51. Bryant Jordan, "Air Force to Allow Enlisted Airmen to Fly Global Hawk Drones," *Military.com*, December 17, 2015, www.military.com/daily-news /2015/12/17/air-force-to-allow-enlisted-airmen-to-fly-global-hawk-drones .html.

52. United States Office of Management and Budget Request FY 2017: 8–21.

53. Hoagland, "Unmanned Air Force," 10.

54. Jordan, "Global Hawk Drones."

55. Andrew Feickert, "U.S. Special Operations Forces (SOF): Background and Issues for Congress," Congressional Research Service, RS21048, April 18, 2016, 1, www.fas.org/sgp/crs/natsec/RS21048.pdf; 2014 Quadrennial Defense Review, 41.

56. 2014 Quadrennial Defense Review, 41.

57. Grace V. Jean, "Equipment Shortages Undercut U.S. Special Operations Forces," *National Defense* magazine, February 2009, www.nationaldefense magazine.org/archive/2009/February/Pages/EquipmentShortagesUnder cutUSSpecialOperationsForces.aspx.

58. "U.S. Special Operations Forces (SOF): Background and Issues for Congress," 6. There are approximately a thousand Navy SEALs in total, since each SEAL

team consists of six SEAL platoons, and each platoon has approximately eighteen sailors.

59. Ibid.

60. 2014 Quadrennial Defense Review, 19.

61. Steven Heffington, "AFPAK to APAC Hands: Lessons Learned," *War on the Rocks*, January 7, 2014, http://warontherocks.com/2014/01/afpak-to-apac -hands-lessons-learned/.

62. Mark W. Lee, "The Afghanistan-Pakistan Hands Program," US Army, February 11, 2014, www.army.mil/article/115523/The_Afghanistan_Pakistan _Hands_Program.

63. General James Muir, USN, "AFPAK Hands Overview," 2012, http://tinyurl .com/hnxoets.

64. Peter Schirmer, Harry J. Thie, Margaret C. Harrell, and Michael S. Tseng, *Challenging Time in DOPMA*, Rand Corporation, 2006, www.rand.org /content/dam/rand/pubs/monographs/2006/RAND_MG451.pdf.

65. Amy Belasco, "Defense Spending and the Budget Control Act Limits," Congressional Research Service, July 22, 2015, www.fas.org/sgp/crs/natsec /R44039.pdf.

66. US Department of Defense, "Fiscal Year 2016 Budget Request Overview," February 2015, 1-1, http://comptroller.defense.gov/Portals/45/Documents /defbudget/fy2016/FY2016_Budget_Request_Overview_Book.pdf.

67. Office of Management and Budget Historical Tables, "Table 5.1: Budget Authority by Function and Subfunction: 1976–2021," 2016.

68. International Institute for Strategic Studies, "The Military Balance, 2016," www.iiss.org/en/publications/military-s-balance.

69. Congressional Budget Office, "Summary of the Bipartisan Budget Act of 2013," December 10, 2013, http://budget.house.gov/uploadedfiles/bba2013summary .pdf.

70. OMB, "Table 5.1."

71. US Department of Defense, "Operation and Maintenance Programs (O-1), Revolving and Management Funds (RF-1), Department of Defense Budget Documents Fiscal Year 2016," February 2015, 145.

72. "A Profile of the Modern Military," in *War and Sacrifice in the Post-9/11 Era*, chap. 6, Pew Research Center, October 5, 2011, www.pewsocialtrends. org/2011/10/05/chapter-6-a-profile-of-the-modern-military/.

73. Defense Travel Management Office. Each individual case will differ based on the legal requirements regarding location, military rank, and family size.

74. "Military Personnel Programs (M-1) Department of Defense Budget Fiscal Year 2017," February 2016, 17.

75. Eric Hanushek, "The Volunteer Military and the Rest of the Iceberg," *Policy Sciences* 8 (1977): 357.

76. See "DOPMA/ROPMA Policy Reference Tool," Rand Corporation, http://dopma-ropma.rand.org/incentives-special-pay.html.

77. Wardynski et al., "Realigning Talent." Also see Tim Kane, "Why Our Best Officers Are Leaving," *The Atlantic*, January/February 2011, www.theatlantic.com/magazine/archive/2011/01/why-our-best-officers-are-leaving/308346/.

78. Hanushek, "The Volunteer Military," 343–61.

79. GAO, "Military Compensation: DOD Needs More Complete and Consistent Data to Assess the Costs and Policies of Relocating Personnel," GAO-15-713, September 9, 2015, 10.

80. DOD, *2014 Demographics: Profile of the Military Community*, 53, http://download.militaryonesource.mil/12038/MOS/Reports/2014-Demographics-Report.pdf.

81. S. Rep. 112–196 (Department of Defense Appropriations Bill, 2013), www.gpo.gov/fdsys/pkg/CRPT-112srpt196/html/CRPT-112srpt196.htm.

82. Kane, *Bleeding Talent.*

83. Russell Golman and Sudeep Bhatia, "Performance Evaluation Inflation and Compression," *Accounting, Organizations and Society* 37, no. 8 (2012): 534–43.

84. Marcus Buckingham, "Trouble with the Curve? Why Microsoft Is Ditching Stack Rankings," *Harvard Business Review* online, November 19, 2013, https://hbr.org/2013/11/dont-rate-your-employees-on-a-curve.

85. Megha Oberoi and Paresh Rajgarhia, "What Your Performance Management System Needs Most," *Gallup Business Journal*, April 4, 2013, www.gallup.com/businessjournal/161546/performance-management-system-needs.aspx.

86. Aina Katsikas, "Same Performance, Better Grades," *The Atlantic*, January 13, 2015.

87. Russell Golman and Sudeep Bhatia, "Performance Evaluation Inflation and Compression," *Accounting, Organizations and Society* 37, no. 8 (2012): 534–43.

88. Kurt Eichenwald, "Microsoft's Lost Decade," *Vanity Fair* 54 (August 2012): 108, www.vanityfair.com/news/business/2012/08/microsoft-lost-mojo-steve-ballmer.

89. Tom Warren, "Microsoft Axes Its Controversial Employee-Ranking System," *The Verge*, November 12, 2013, www.theverge.com/2013/11/12/5094864/microsoft-kills-stack-ranking-internal-structure.

90. Max Nisen, "Why GE Had to Kill Its Annual Performance Reviews after More Than Three Decades," *Quartz*, August 13, 2015, http://qz.com/428813/ge-performance-review-strategy-shift/.

91. Drake Baer, "Why Adobe Abolished the Annual Performance Review and You Should, Too," *Business Insider,* April 10, 2014, www.businessinsider.com /adobe-abolished-annual-performance-review-2014-4.

92. Ibid.

93. Deidra J. Schleicher, Rebecca A. Bull, and Stephen G. Green, "Rater Reactions to Forced Distribution Rating Systems," *Journal of Management* 35, no. 4 (August 2009).

94. American National Standards Institute, Inc., *Performance Management Standard,* November 30, 2012, www.shrm.org/resourcesandtools/tools-and -samples/how-to-guides/documents/12-0794%20performance%20 mngmt%20standard_interior_viewonlyfnl_rvsd10-4-13.pdf.

95. Stephen Losey, "New EPR System: Leaders Answer Your Questions," *Air Force Times,* January 12, 2015, www.airforcetimes.com/story/military/careers /air-force/2015/01/12/new-epr-system/21460541/.

96. For an example of an Officer Evaluation Report (in this case for the Coast Guard), go to www.uscg.mil/forms/oer/CG_5310B.pdf.

97. Stephane L. Wolfgeher, "Inflation of USAF Officer Performance Reports: Analyzing the Organizational Environment" (master's thesis, Naval Postgraduate School, Monterey, 2009).

98. Herman Aguinis and Ernest O'Boyle, Jr., "The Best and the Rest: Revisiting the Norm of Normality of Individual Performance," *Personnel Psychology* 65, no. 1 (Spring 2012): 79–119.

99. Basketball Reference, www.basketball-reference.com/leagues/NBA_1982 _totals.html.

100. Marcus Buckingham and Ashley Goodall, "Reinventing Performance Management," *Harvard Business Review,* April 2015.

101. Timothy R. Reese, "Transforming the Officer Evaluation System: Using a 360-Degree Feedback Model," US Army War College, April 2002.

102. Chaitra M. Hardison, Mikhail Zaydman, Oluwatobi A. Oluwatola, Anna Rosefsky Saavedra, Thomas Bush, Heather Pete, and Susan G. Straus, *360-Degree Assessments: Are They the Right Tool for the U.S. Military?* (Santa Monica, CA: Rand Corporation, 2015), www.rand.org/pubs/research_reports /RR998.html.

103. Bernard D. Rostker, Harry Thie, James Lacy, Jennifer Kawata, and Susanna Purnell, *The Defense Officer Personnel Management Act of 1980: A Retrospective Assessment* (Santa Monica, CA: Rand Corporation, 1993), 79, www.rand .org/pubs/reports/R4246.html.

104. Ibid.

105. Legal Information Institute, "10 U.S. Code § 619—Eligibility for Consideration for Promotion: Time-in-Grade and Other Requirements," www.law.cornell.edu/uscode/text/10/619.

106. Flag officers face a different process. For those higher ranks, promotions are made by the president with the "advice and consent" of the Senate. US Code 10 § 526 limits the number of active-duty general and flag-rank officers allowed. Specifically, the Army may not exceed 231 general officers ranking brigadier general, major general, lieutenant general, and general. The Air Force is capped at 198 officers ranking brigadier general or above. The Marines can only have 61 general officers. The Navy is limited to no more than 162 officers ranking rear admiral (upper), vice admiral, and admiral.

107. 10 US Code § 624: Promotions: how made (a) (1). A promotion list of active-duty, all-fully-qualified officers is approved by the president, at the same time, and treated the same way within chapter 36 of US Code 10. This list is only created when the secretary of defense confirms that all officers on the list are needed for the next higher grade, usually in order to accomplish a specific mission or goal. Officers on the all-fully-qualified list are qualified and meet the criteria to be promoted by a selection board.

108. AFI 36-2501, 15.

109. 10 US Code § 624: Promotions: how made (a).

110. Stephen Losey, "514 Officers Tapped for Squadron Command," *Air Force Times*, December 9, 2015, www.airforcetimes.com/articles/514-officers-tapped-for-squadron-command.

111. See Kane, "The Leader/Talent Matrix." This chapter does not include details from the working paper that contrast the armed forces with other private- and public-sector organizations.

112. Andrew Tilghman, "White House Cancels Next Pay Review," *Military Times*, January 9, 2015, www.militarytimes.com/story/military/benefits/pay/2015/01/09/no-qrmc-obama/21515187/.

113. Rostker, *I Want You!* 719.

114. Todd Harrison, "Rebalancing Military Compensation: An Evidence-Based Approach," Center for Strategic and Budgetary Assessment, July 12, 2012, http://csbaonline.org/publications/2012/07/rebalancing-military-compensation-an-evidence-based-approach/.

115. See "Statistical Report on the Military Retirement System Fiscal Year 2011," Department of Defense, Office of the Actuary, May 2012, http://actuary.defense.gov/Portals/15/Documents/statbook11.pdf.

116. Mark Cancian, "The Impact of Rising Compensation Costs on Force Structure," *Joint Forces Quarterly* 79, no. 4 (2015), http://ndupress.ndu.edu /Portals/68/Documents/jfq/jfq-79/jfq-79_77-82_Cancian.pdf.

117. GAO, "Military Personnel: Military and Civilian Pay Comparisons Present Challenges and Are One of Many Tools in Assessing Compensation" GAO-10-561R, April 1, 2010, www.gao.gov/assets/100/96645.pdf.

118. Read more at US Navy, "1794 Navy and Marine Corps Ration Chart," www.navycs.com/charts/1794-navy-rations.html.

119. Defense Finance and Accounting Services, "Special and Incentive Pays," September 12, 2014, www.dfas.mil/militarymembers/payentitlements /specialpay.html.

120. See US Navy, "Navy Enlisted Pay Chart 1893," www.navycs.com/charts /1893-navy-pay-chart.html.

121. Beth J. Asch, Paul Heaton, James Hosek, Paco Martorell, Curtis Simon, and John T. Warner, "Cash Incentives and Military Enlistment, Attrition, and Reenlistment," Rand Corporation, 2010, www.rand.org/pubs/monographs /MG950.html.

122. Report of the Military Compensation and Retirement Modernization Commission, January 29, 2015, 2, www.mcrmc-research.us/02%20-%20 Final%20Report/MCRMC-FinalReport-29JAN15-HI.pdf.

123. Ibid., 12.

124. The stream of future pension payments after a military retirement will vary based on an individual's lifespan and future cost-of-living adjustments (COLAs) to reflect price inflation. If we assume the typical enlisted retiree lives for forty-three more years and earns $28,748 per year in retirement pay, then the total retirement amount is $1.24 million (not counting cost-of-living adjustments).

Index

About the Author

Tim Kane is the JP Conte Fellow in Immigration Studies at the Hoover Institution at Stanford University. A graduate of the US Air Force Academy, he served as a US Air Force intelligence officer with two tours of overseas duty in Seoul, South Korea, and Tokyo, Japan.